U S Army CGSC Space Reference Text
2012

Developed by

US Army Command and General Staff College

In Conjunction with

US Army Space and Missile Defense Command

Submit all recommended changes to:

Joint Space Education Division

Department of Joint, Interagency and Multinational Operations

US Army Command and General Staff College

Attn: Mr. Thomas A. Gray

100 Stimson Ave, Room 4542

Fort Leavenworth, Kansas 66027

Phone: DSN 552-2799; Comm: (913) 684-2799

E-Mail: thomas-gray@us.army.mil

E-Mail: cliff.hodges@us.army.mil

E-Mail: jason.b.burch.mil@mail.mil

2012 Update Project Team:

Mr Tom Gray, SMDC LNO

LTC Cliff Hodges, USA

MAJ Jason Burch, USAF

Joint Space Operations

US Army Command and General Staff College

DISCLAIMER

Opinions, conclusions, and recommendation expressed or implied within are solely those of the authors and do not necessarily represent the views of the United States Army Command and General Staff College, the U.S. Army, the Department of Defense, or any other U.S. government agency.

US Army CGSC Space Reference Text
Table of Contents

PURPOSE

The purpose of this reference text is to provide information about space systems and their use for US Army and joint military operations. The intended primary users are US Army Command and General Staff College students, although military commanders, staff officers, Noncommissioned Officers, instructors, and anyone else applying space-based systems for military purposes may also find it useful. Military space operations are truly joint in their nature. There are many references to policies, programs, operations, and applications from many US Government organizations, the military services, and commercial entities.

The format and contents provide a central reference regarding the environment of space, the capabilities and limitations of US satellite systems, space applications that support the military strategic, operational, and tactical levels, and space support planning considerations. In addition, the text provides a cursory background in space law, space policy, orbital mechanics, and threats to US dominance of space.

SCOPE

The scope of this book provides a fundamental knowledge of space and space systems. The intent is for this to be a single reference source. Primary information is on space systems used by the US military. Because of the proliferation of space systems, information about other nations' systems is included on a limited basis. This reference excludes classified information to allow wider dissemination and access.

OVERVIEW

Space capabilities permeate virtually every military mission area. Space is no longer the frontier battleground of some future time frame. It is a medium vital to successful military operations around the globe, and is recognized as one of the domains of warfare (i.e., land, sea, air, cyber, and space) through and from which Army operations may be conducted globally now and into the future. Space is a distinct operating environment that is different from land, air, and sea domains. Space-based assets transcend geographical borders unimpeded. Since there are no recognized political boundaries in space, satellites take advantage "open skies" global coverage.

As the Army continues to evolve as a 21st Century warfighting force, soldiers must be trained to understand the potential benefits of combat multipliers derived from space assets and how to use them effectively. The medium of space and space products are increasingly critical considerations for leaders and planners at all levels. Space-based systems and capabilities provide near to real time relevant information directly to the warfighter and quickly "link" the military force into mission command, sensor, and weapon systems network, which integrates and orchestrates effects within the joint operational environment. Space capabilities dramatically integrate the effects of widely dispersed weapons platforms and forces, provide dominant awareness of the battlefield, and allow for precision engagement and dominant maneuver. Access to and the availability of space capabilities, as well as awareness of their limitations, is critical for unified land operations. Understanding the key aspects of space operations enables military and DoD personnel to effectively integrate space capabilities into military options.

Space is the ultimate high ground today and will be in the future. Depending upon redundancies and mix of payloads, satellites can provide capabilities day/night and in all weather. Operation DESERT

STORM in 1991 proved that space systems were vital assets to enhance land warfare operations. This was the first occasion when the full range of modern military space assets was applied to terrestrial conflict at a tactical and operational level. Space systems provided the communications infrastructure to command and control mobile armor formations and to synchronize precision strikes. The Global Positioning System (GPS) allowed precision navigation and precision strikes across the featureless Arabian Peninsula, and the Defense Support Program (DSP) satellites warned Army forward deployed forces of imminent ballistic missile attack. Military operations are evolving as integral space-based assets provide relevant and time-sensitive information directly to the warfighter. This is evident more so today as seen through the implementation and integration of commercial imagery, environmental monitoring, the increased use of SATCOM, dependency on information to support decisive action.

The expansion of new players in the military space arena may prove to be one of the most distinguishing features for future warfare. Almost every country has access to communication satellites. Multispectral and Synthetic Aperture Radar imagery is commercially available. Current weather pictures from satellites are available worldwide. Many nations have some degree of indigenous capability to employ militarily useful satellites today and more are expected to take advantage of those capabilities in the future.

Military and non-military activities in space are increasingly interdependent and terrestrial military forces are becoming ever more dependent on an ever-growing number of operational military spacecraft. The division between military and non-military space activities has never been clear, and is becoming increasingly blurred. Some have made the analogy that space is to the Information Age as oil was to the Industrial Age. The control and protection of national, civil and commercial space systems will become paramount to national and international security in this century. Growing military and non-military dependence upon space capabilities is recognized by our adversaries and could eventually increase the risk to those systems.

Chapter 1
Space Law, Policy, and Doctrine

OVERVIEW

Space policy, law and doctrine define the goals and principles of the U.S. space program. International and domestic laws and regulations, national interests and security objectives shape the US space program. Furthermore, fiscal considerations both shape and constrain space policy. Space policy formulation is a critical element of the US national planning process, as it provides the framework for future system requirements. This chapter examines the international and domestic legal parameters within which the US must conduct its space programs. It also outlines the basic tenets of US policy and doctrine.

Space policy continues to evolve based on revised goals and objectives of the nation, budget constraints, previous space policies, current programs, national and international law, and treaty obligations. DoD Space Policy and the Army Space Policy support the National Space Policy.

The current National Space Policy and several other US Government space related policies can be found at www.globalsecurity.org/space/library/policy/national.

INTERNATIONAL LAW

International law consists of agreements and customary law as adopted by nations.

International agreements are treaties, conventions, protocols, or Executive Agreements that have been signed and/or adopted by various nations and are either bi-lateral (between two countries) such as the former Anti-Ballistic Missile (ABM) Treaty that was between the Russia and the USA, or multi-lateral (between numerous nations).

Customary law derives from the practice of nations over time. Certain conduct and customs implemented continuously over time eventually become the standard expected of all nations in particular situations. "When such a practice attains a degree of regularity and is accompanied by the general conviction among nations, that behavior in conformity with that practice is obligatory, it can be said to have become a rule of customary law binding upon all nations."[1] Often customary law once widely adopted is formalized in a written document such as a treaty, convention, or protocol. Many agreements, adopted and signed by a majority of nations, are considered customary international law which all nations, whether signatories or not, become obligated to follow upon an agreement's overwhelming acceptance in the international community. In many instances where the US is not a signatory, or opposes the inferred customary obligation on the grounds that a particular agreement is not widely accepted, then there is not obligation to adhere to that custom.

---

[1] Thomas, A.R. and Duncan, James C. "Annotated Supplement to the Commander's Handbook on the Law of Naval Operations," US Naval War College International Law Studies, Vol. 73, Newport, Rhode Island 1999.

Why does the U.S., the only remaining superpower, choose to or have to follow international agreements?   Once the U.S. ratifies a treaty or agreement, then it becomes U.S. law. After discussions among nations regarding an international agreement, the President of the United States either signs or rejects the agreement. He may sign with reservations as to certain clauses as well. The agreement is then presented to the Senate. Two thirds of the Senate must give its advice and consent before the President can ratify the agreement. Once ratified, the agreement then becomes U.S. law. Article II of the U.S. Constitution states, "[the President] shall have Power, by and with the Advice and Consent of the Senate to make Treaties . . ." Article VI states, "[t]his Constitution, and the laws of the United States which shall be made in Pursuance thereof; and all Treaties made, or which shall be made, under the Authority of the United States, shall be the supreme law of the land".

The U.S. becomes a party to selected international laws and agreements because they make good sense, such as reducing nuclear arms and chemical weapons, and they help to further our foreign and domestic policies as well as strengthen our national security situation. One example is abolition of the use of chemical weapons. The use of chemical weapons is cruel due to its mass casualty effects and the inability to contain it and prevent civilian casualties. The U.S. has ratified the Chemical Weapons Convention (CWC), which prohibits the stockpiling and use of chemical weapons.   Often the U.S. decides to become a party to a treaty but does not agree to every article or principles of the treaty, and therefore may sign and/or ratify with reservations. For instance, the U.S. voiced at least one reservation with the CWC. The U.S. reserved the right to use riot control agents (RCA) such as those utilized by local police. Use of RCA requires President of the United States and Secretary of Defense approval since the dispersal of any type of chemical, even harmless gases could be interpreted as a chemical attack. Additionally, the U.S. refuses to sign many treaties. For instance the U.S. did not sign or ratify the Moon Treaty. Although it contains some favorable principles, most of them benefited non-space faring nations to the detriment of the space faring nations. The favorable principles were already present in the Outer Space Treaty to which the U.S. is a party.

A final complaint about international law which has been echoed many times is that other countries, such as North Korea, do not always follow international law, so what is the U.S. incentive? First of all, the majority of nations do respect international law. Nations do not normally become a party to a treaty or agreement unless it confers upon them some benefit. The treaties and agreements the U.S. has become a party to provide at least some benefit. Especially in the face of adversary noncompliance, continued U.S. compliance with international law facilitates international and public support for U.S. conduct. A major issue regarding the U.S. reluctance to back out of the ABM Treaty was our reliance on other strategic arms limitations treaties, which Russia said would weaken by the U.S. withdrawal from the ABM Treaty. The U.S. also enters some treaties because they address what the U.S. considers morally correct conduct (e.g., the POW treaty). Adversary noncompliance should not affect America's moral stance.

SPACE LAW

The term *space law* refers to a body of law drawn from a variety of sources and consisting of two basic types of law: international and domestic. The domestic portion is legislation passed by Congress and regulations promulgated by executive agencies of the US government.

Table 1-1 at the end of the chapter summarizes, by date, key international agreements and US domestic laws that affect the scope and character of US military space activities. Listed below are some of the more important basic principles and rules.

The Outer Space Treaty (or the Treaty on Principles Governing the Activities of States in the Exploration and Use of Outer Space Including the Moon and Other Celestial Bodies) is the main treaty relating to the use of outer space and celestial bodies; is the backbone for most other agreements and laws relating to space; and guides how nations conduct activities in space. The 1967 Outer Space Treaty has been called the Magna Carta of space and includes approximately 126 nations as signatories, many of which are not space faring nations. This treaty is a quite amazing document considering the date it was conceived, how little was known about outer space, and the fact that very little space activity existed at the time. Most of the principles discussed below derive from the Outer Space Treaty (OST).

Principles

One principle which is worthy of noting prior to discussing space law principles is the fact that treaties and agreements, unless specifically stated within their text, do not apply to belligerents in time of war or international conflict. During a war the Outer Space Treaty therefore does not apply between the belligerents; however, this principle does not then give belligerents carte blanche to ignore all treaties and agreements. Most if not all treaties and agreements, certainly multilateral documents, continue to apply between the belligerents and other non-belligerent signatories. Belligerents must continue to respect the rights of other nonbelligerent nations in outer space.

International law, and specifically the United Nations (UN) Charter, applies to outer space (OST Art. I and III). This causes the principles of the UN Charter and all other international law to be applicable to outer space. The UN Charter requires all members to settle disputes by peaceful means, and prohibits the threat to use or actual use of force against the territorial integrity or political independence of another state.

The Outer Space Treaty prohibits claims of sovereignty over outer space and celestial bodies. Threats or use of force can occur from space to earth as well as to nations' property in space such as satellites. The US has declared their satellites to be sovereign property, which if attacked or interfered with will be defended. This is in line with Article VIII of the Outer Space Treaty which states that the launching State of an object in space retains jurisdiction over that object and is liable for damage caused by that object (see also the Liability Convention).

The UN Charter also recognizes a state's inherent right to act in individual or collective self-defense.

Outer space, the Moon, and other celestial bodies are not subject to appropriation by claim of sovereignty, use or occupation, or any other means (OST Art. II) In 1976, eight equatorial countries claimed sovereignty over the geostationary orbital arc above their territory. Most other countries, including all major space powers, rejected the claim. As briefly discussed above, nations such as the US have claimed satellites and other space objects as their sovereign property, protected against interference and attack similar to protections afforded flagged vessels in international waters.

Outer space is free for use by all countries. This principle relates to the non-appropriation principle and is analogous to the right of innocent passage on the high seas.

Outer space will be used for peaceful purposes only. Most Western nations, including the US, equate peaceful purposes with non-aggressive ones. Consequently, all non-aggressive military use of space is permissible, except for specific prohibitions of certain activities noted elsewhere in this section. Initially, the Soviet Union interpreted "peaceful purposes" to mean no military. The US has always maintained that "peaceful purposes" means no aggressive military actions. This begs the question, how do you interpret "peaceful purposes," is it based on a universal agreement among all parties to the Outer Space Treaty or left to each nation individually? In order to engage in aggressive acts in space does a nation need the blessing of the UN Security Council, or can a nation just claim they are acting in self-defense?

Objects launched into space must be registered with the UN. (Registration Convention) All objects launched into outer space must be placed on the registration document of a nation, and this document must then be filed with the United Nations. Typically it is the launching nation that registers a satellite, but in some cases one nation launches for another and the registration is of the owner. When an international commercial consortium with no particular nation identification owns the satellite and it is launched in international waters by a commercial company, that company should still register that satellite; however, it is neither a signatory nor a nation state. There may be instances where not every satellite is then registered.

A country retains jurisdiction and control over its registered space objects. (OST Art. VIII) This rule applies regardless of the condition of the objects. Whether the object is orbiting space junk or has deorbitted and landed in another country, the registered owner is still responsible.

A country is responsible for regulating, and is ultimately liable for, the outer space activities of its citizens. (OST Art. VII) This article of the Outer Space Treaty and more specifically the Liability Convention cause registered nations to be absolutely liable for all space objects that deorbit and cause damage in the atmosphere and on the earth. A type of shared liability applies when objects collide in outer space.

Nuclear weapons tests and other nuclear explosions in outer space are prohibited. (Limited Test Ban Treaty of 1963) Before this prohibition in 1963, the US exploded three small nuclear devices in outer space in Project Argus. This occurred over a period of two weeks; such an experiment would not be permissible today.

Nuclear weapons and other weapons of mass destruction (such as chemical and biological weapons) may not be placed into orbit, installed on celestial bodies, or stationed in space in any other manner. (OST Art. IV) Notice that the ban on weapons in space is only against nuclear weapons and weapons of mass destruction which includes nuclear, biological, and chemical weapons. There is no prohibition against having a nuclear powered satellite or against having other weapons in space.

A country may not test any kind of weapon; establish military bases, installations, or fortifications, nor conduct military maneuvers on celestial bodies. (OST Art. IV) This reinforces the principles that nations may not claim sovereignty over outer space and celestial bodies, that space is to be used for peaceful purposes (non-aggressive military), and outer space free for use by all.

Interfering with national technical means of verification is prohibited provided such systems are operating in accordance with generally recognized principles of international law and are in fact being used to verify provisions of specific treaties.

The point here is to not accidentally precipitate a war by interfering with a nation's ballistic missile attack warning system. These systems reduce the likelihood of accidental nuclear war by confirming that the other side has not launched ballistic missiles. This removes doubt about an adversary's actions.

Any act not specifically prohibited is permissible. Thus, even though the list (see **Table 1-1**) of prohibited acts is sizable, there are few legal restrictions on the use of space for non-aggressive military purposes. As a result, international law implicitly permits the performance of such traditional military functions as surveillance, reconnaissance, navigation, meteorology and communications. It permits space stations along with testing and deployment in Earth orbit of non-nuclear weapon systems. This includes anti-satellite weapons; space-to-ground conventional weapons; the use of space for individual and collective self-defense; and any conceivable activity not specifically prohibited or otherwise constrained.

DOMESTIC SPACE LAW

Domestic law has always shaped military space activities through the spending authorization and budget appropriation process. For example, Congress deleted funding for further testing of the USAF's direct ascent ASAT weapon in the mid- 1980s, and thereby canceled the program. In addition, a number of laws not designed solely to address space have a space aspect. For instance, under the Communications Act of 1934, the president has the authority to gain control of private communications assets owned by US corporations during times of crisis. Since the 1960s, this authority has included both the terrestrial and space segments of domestically owned communications satellites. Space-specific legislation (beyond the annual National Aeronautics and Space Administration (NASA) authorization) is a relatively recent activity.

The Reagan administration placed emphasis on the creation of a third sector of space activity, commercial space, in addition to the traditional military and civil sectors. For example, Congress passed the Commercial Space Launch Act of 1984 to facilitate the development of a commercial launch industry in the U.S. From a DoD perspective, the importance of this legislation lies in its authorization for commercial customers to use DoD launch facilities on a reimbursable basis. Thus, the DoD is in the business of overseeing commercial operations from its facilities and placing commercial payloads in the launch queue. Although a recent development, there is a trend towards intertwining the commercial space industry and DoD space programs. This is seen even more so with the current National Space Policy of June 2010.

One domestic law that clearly applies to potential space related operations is 18 United States Code (USC) 1367, the infamous "Captain Midnight" law so named after a US citizen, John R. McDougall, gained access to a U.S. television satellite and inserted his own broadcast on cable station HBO protesting the scrambling of the signal and paying for it. The law prohibits the unauthorized harmful interference with a U.S. satellite.

NATIONAL SPACE POLICY

A nation's space policy is extremely important, especially as it relates to space law and space doctrine. In order to understand present U.S. space policy and attempt to predict its future, an examination of its evolution is necessary. The national space policy contains guidelines and actions to implement in the conduct of space programs and related activities. The present United States space policy is very important in understanding how future space strategies may evolve. Keep in mind, that while policy provides space goals and a national framework, national interests and national security objectives actually shape the policy. This framework will lead towards building and meeting future U.S. requirements and subsequent national space strategies.

Early Policy

The launch of Sputnik I on 4 October 1957 had an immediate and dramatic impact on the formulation of U.S. space policy. Although the military had expressed an interest in space technology as early as the mid 1940s, a viable program failed to emerge for several reasons. These include intense inter-service rivalry, military preoccupation with the development of ballistic missiles that prevented a sufficiently high funding priority from being assigned to proposed space systems, and national leadership that did not initially appreciate the strategic and international implications of emerging satellite technology. Once national leadership gained this appreciation, it became committed to an open and a purely scientific space program.

The emergence of Sputnik I transposed this line of thought: besides clearly demonstrating the Soviets had the missile technology to deliver payloads at global ranges, Sputnik led to much wider appreciation of orbital possibilities. The result was the first official U.S. government statement that space indeed, was of military significance. This statement, issued on 26 March 1958 by President Dwight D. Eisenhower's science advisory committee, stated that the development of space technology and the maintenance of national prestige were important for the defense of the U. S. Congress also quickly recognized that space activities were potentially vital to national security.

The first official national space policy was the National Aeronautics and Space Act (NASA) of 1958. This Act stated the policy of the United States was to devote space activities to peaceful purposes for the benefit of all humankind. It mandated separate civilian and national security space programs and created a new agency (NASA) to direct and control all US space activities, except those "peculiar to or primarily associated with the development of weapons systems, military operations, or the defense of the United States." The Department of Defense was to be responsible for these latter activities.

A legislative basis for DoD responsibilities in space was thereby provided early in the space age. The act established a mechanism for coordinating and integrating military and civilian research and development. It also encouraged significant international cooperation in space and called for preserving the role of the US as a leader in space technology and its application. Thus, the policy framework for a viable space program was in place. The principles enunciated by NASA have become basic tenets of the US space program. These tenets included: peaceful focus on the use of space, separation of civilian and military space activities, emphasis on international cooperation and preservation of a space role. All presidential space directives issued since 1958 have reaffirmed these basic tenets.

However, a space program of substance still did not exist. The Eisenhower administration's approach to implementing the new space policy was conservative, cautious and constrained. The government consistently disapproved of the early DoD and NASA plans for manned space flight programs. Instead the administration preferred to concentrate on unmanned, largely scientific missions and to proceed with those missions at a measured pace. It was left to subsequent administrations to give the policy substance.

Intervening Years

Two presidential announcements, one by John F. Kennedy on 25 March 1961 and the second by Richard M. Nixon on 7 March 1970, were instrumental in providing the focus for the U.S. space program. The Soviet Union launched and successfully recovered the world's first cosmonaut. Although Yuri Gagarin spent just 89 minutes in orbit, his accomplishment electrified the world.

The American response articulated by President Kennedy as a national challenge to land a man on the Moon and return him safely to Earth defined US space goals for the remainder of the decade. The Kennedy statement came during a period of intense national introspection. Prestige and international leadership were clearly the main objectives of the Kennedy space program. However, the generous funding that accompanied the Apollo program had important collateral benefits as well. It permitted the buildup of U.S. space technology and the establishment of an across-the-board space capability that included planetary exploration, scientific endeavors, commercial applications and military support systems.

President Johnson's years in office saw the commencement of work on nuclear ASATs and the implementation of several pertinent treaties to include the Outer Space Treaty, the Astronaut Rescue and Return Agreement, and the Nuclear Test Ban Treaty.

As the 1960s drew to a close, a combination of factors including domestic unrest, an unpopular foreign war and inflationary pressures forced the nation to reassess the importance of the space program. Against this backdrop, President Nixon made his long-awaited space policy announcement in March 1970. His announcement was a carefully considered and worded statement that was clearly aware of political realities and the mood of Congress and the public. In part, it stated:

"Space expenditures must take their proper place within a rigorous system of national priorities....Operations in space from here on in must become a normal and regular part of national life. Therefore, they must be planned in conjunction with all of the other undertakings important to us."

Although spectacular lunar and planetary voyages continued until 1975 as a result of budgetary decisions made during the 1960s, the Nixon administration considered the space program of intermediate priority and could not justify increased investment or the initiation of large new projects. One major notable event of this time is that Nixon approved the development of the space shuttle.

Within the DoD, this accentuation on practicality translated into reduced emphasis on manned spaceflight, but led to the initial operating capability for many of the space missions performed today. For example, initial versions of the systems were all developed and fielded during this period. These versions are now known as the Defense Satellite Communications System, the Defense Support Program, the Defense Meteorological Satellite Program and the Navy's Transit Navigation Satellite Program (later to evolve as the Global Positioning System).

Carter Administration Space Policy

President Jimmy Carter's administration conducted a series of interdepartmental studies to address the malaise that had befallen the nation's space effort. The studies addressed apparent fragmentation and possible redundancy among civil and national security sectors of the U.S. space program. It also sought to develop a coherent recommendation for a new national space policy. These efforts resulted in two 1978 presidential directives (PD): PD-37, on national space policy and PD-42, on civil space policy.

PD-37 reaffirmed the basic policy principles contained in the National Aeronautics and Space Act of 1958. It identified the broad objectives of the U.S. space program, including the specific guidelines governing civil and national security space activities.

PD-37 was important from a military perspective because it contained the initial, tentative indications that a shift was occurring in the national security establishment's view on space. Traditionally, the military had seen space as a force enhancer or an environment in which to deploy systems to increase the effectiveness of land, sea, and air forces. Although the focus of the Carter policy was clearly on restricting the use of weapons in space, PD-37 reflected an appreciation of the importance of space systems to national survival; a recognition of the Soviet threat to those systems; and a willingness to push ahead with development of an anti-satellite capability in the absence of verifiable and comprehensive international agreements restricting such systems. In other words, the administration was beginning to view space as a potential war-fighting medium.

PD-42 was directed exclusively at the civil space sector to guide U.S. efforts over the next decade. However, it was devoid of any long-term space goals, expecting the nation to pursue a balanced evolutionary strategy of space applications, space science, and exploration activities. The absence of a more visionary policy reflected the continuing developmental problems with the shuttle and the resulting commitment of larger than expected resources.

Reagan Administration Space Policy

President Ronald Reagan's administration published comprehensive space policy statements in 1982 and 1988. The first, pronounced on 4 July 1982 and embodied in National Security Decision Directive 42 (NSDD-42), reaffirmed the basic tenets of previous (Carter) U.S. space policy. It also placed considerable emphasis on the STS as the primary space launch system for both national security and civil government missions. In addition, it introduced the basic goals of promoting and expanding the investment and involvement of the private sector in space.

The single statement of national policy from this period that most influenced military space activities and illuminated the transition to a potential space warfighting framework is NSDD-85, dated 25 March 1983. Within this document, President Reagan stated his long-term objective to eliminate the threat of nuclear-armed ballistic missiles through the creation of strategic defensive forces. This NSDD coincided with the establishment of the Strategic Defense Initiative Organization (SDIO) and represented a significant step in the evolution of U.S. space policy. Since 1958, the U.S. had, for a variety of reasons, refrained from crossing an imaginary line from space systems designed to operate as force enhancers to establishing a warfighting capability in space. The anti-satellite (ASAT) initiative of the Carter administration was a narrow response to a specific Soviet threat. However, the SDI program represented a significant expansion in the DoD's assigned role in the space arena.

The second comprehensive national space policy incorporated the results of a number of developments that had occurred since 1982, notably the U.S. commitment in 1984 to build a space station and the space shuttle Challenger.

For the first time, the national space program viewed commercial space equal to the traditional national security and civil space sectors. Moreover, the new policy dramatically retreated from its previous dependence on the STS and injected new life into expendable launch vehicle programs. In the national security sector, this program was the first to address space control and force application at length, further developing the transition to warfighting capabilities in space.

In 1988, the last year of the Reagan presidency, Congress passed a law allowing creation of a National Space Council (NSPC), a cabinet-level organization designed to coordinate national policy among the three space sectors. The incoming administration would officially establish and very effectively use the National Space Council.

Bush Administration Space Policy

Released in November 1989 as National Security Directive 30 (NSD-30), and updated in a 5 September 1990 supplement, the Bush administration's national space policy retained the goals and emphasis of the final Reagan administration policy. The Bush policy resulted from an NSPC review to clarify, strengthen and streamline space policy, and has been further enhanced by a series of national space policy directives (NSPD) on various topics. Areas most affected by the body of Bush policy documentation included:

- U.S. Commercial Space Policy Guidelines

- Provision of a framework for the National Space Launch Strategy

- Landsat Remote Sensing Strategy

The policy reaffirmed the organization of U.S. space activities into three complementary sectors: civil, national security, and commercial. The three sectors coordinate their activities to ensure maximum information exchange and minimum duplication of effort.

Space leadership is a fundamental objective guiding U.S. space activity. The policy recognized that leadership does not require preeminence in all areas and disciplines of space operations but does require U.S. preeminence in those key areas critical to achieving space goals:

- To strengthen the security of the United States;

- To obtain scientific, technological, and economic benefits for the general population and to improve the quality of life on Earth through space-related activities;

- To encourage continuing United States private sector investment in space and related activities;

- To promote international cooperative activities, taking into account United States national security, foreign policy, scientific, and economic interests;

- To cooperate with other nations in maintaining the freedom of space for all activities that enhance the security and welfare of mankind; and

- To expand human presence and activity beyond Earth orbit into the solar system.

These general goals were not much different from those articulated by President Carter in 1978, and their heritage went back as far as the 1958 National Aeronautics and Space Act. The major changes were: increasing detail in policy objectives and implementation guidelines; the introduction and expansion of emphasis on commercial space activities; and a maturing recognition that space, like land, sea, and air, is a potential warfighting medium.

The Bush policy detailed specific policy, implementing guidelines, and actions for each of the three space sectors and inter-sector activities. The civil sector was to engage in all manners of space-related scientific research, develop space-related technologies for government and commercial applications, and establish a permanent manned presence in space with NASA as the lead civil space agency.

Commercial policy centered on government activities to promote and encourage commercial space-related endeavors. These efforts sought to secure economical and other benefits to the nation that a healthy and vigorous commercial space industry would provide. NASA and the Departments of Defense, Commerce, and Transportation were to work cooperatively with the commercial sector to make government facilities and hardware available on a reimbursable basis.

The U.S. will conduct those activities in space that are necessary to national defense. Such activities contribute to security objectives by: (1) deterring or, if necessary, defending against enemy attack; (2) assuring that enemy forces cannot prevent our use of space; (3) negating, if necessary, hostile space systems; and (4) enhancing operations of U.S. and allied forces. In order to accomplish these objectives, DoD was to develop, operate and maintain a robust space force structure capable of satisfying the mission requirements of space support, force enhancement, space control and force application.

Clinton Administration Space Policy

A repositioning of priorities in the Clinton Administration was reflected by the decision in August 1993, to merge various White House science and technology councils into one National Science and Technology Council (NSTC), which would do most of the day-to-day work through permanent or ad hoc interagency working groups. The National Space Council was absorbed into the new "NSTC" along with the National Critical Materials Council and the Federal Coordinating Council for Science, Engineering and Technology.

The White House structure for articulating National policy for science and technology was put in place by the Presidential Review Directive (PRD)/NSTC series and the Presidential Decision Directive (PDD)/NSTC series as established by PDD/NSTC 1. Within four months during the summer of 1994, three additional policies were established articulating Clinton's space policy.

PDD/NSTC 2 called for the Department of Commerce and Defense "to integrate their programs into a single, converged, national polar-orbiting operational environmental satellite system." This began occurring in 1997. The DMSP satellite program is to merge with the National Oceanic Atmospheric Administration (NOAA) satellite program. The new system formed by the merger of the two programs would be known as the National Polar Orbiting Environmental Satellite (NPOES) System. The DMSP merge was expected to be completed by September 1998; however, it has since been rescinded as a program by the Obama administraion.

PDD/NSTC 3 (replaced Bush's NSPD 5 designed to continue the LANDSAT program) assures "the continuity of LANDSAT-type and quality of data," and reduces the "risk of data gap."

PDD/NSTC 4 superseded all previous policies for US space transportation and "establishes national policy, guidelines, and implementation actions for the conduct of national space transportation programs." It also allocated space transportation responsibilities among Federal civil and military agencies.

In September 1996, the Clinton administration released a new National Space Policy. The new Policy had five goals:

- Knowledge by exploration(1989)

- Maintain national security (1989)

- Enhance competitiveness and capabilities (new)

- Private sector investment (1989)

- Promote international cooperation (1989)

These goals were very similar to those established in 1978 by President Carter and their heritage goes back as far as the 1958 National Aeronautics and Space Act under Eisenhower. Of these five goals, the third was new from the 1989 National Space Policy.

For each major area covered in the 1996 National Space Policy (Civil space, Defense space, Intelligence space, Commercial space and Intersector space), a set of guidelines similar to the ones in the 1989 National Space Policy was established.

Civil Space Sector Guidelines. NASA is designated as the lead agency and is tasked with the following:

- Develop and operate the International Space Station (ISS) to support activities requiring the unique attributes of humans in space and establish a permanent human presence in Earth orbit.

- Work with the private sector to develop flight demonstrators that support a decision by the end of the decade on development of a next-generation reusable launch system (RLS).

- Undertake a sustained program to support a robotic presence on the surface of Mars by the year 2000 for the purposes of scientific research, exploration and technology development.

National Security Space Sector Guidelines. DoD is the lead agency for this sector. This sector is divided into defense guidelines and intelligence guidelines. First - defense guidelines:

- Improve support to military operations.

- DoD will maintain the capability to execute the mission areas of space support, force enhancement, space control and force application.

- DoD will be the lead agency for improvement and evolution of the current expendable launch vehicle (ELVs) fleet, including appropriate technology development.

- DoD will pursue integrated satellite control and continue to enhance the robustness of its satellite control capability.

- Consistent with treaty obligations, the U.S. will develop, operate and maintain space control capabilities to ensure freedom of action in space and, if directed, deny such freedom of action to adversaries.

- The US will pursue a ballistic missile defense program to provide for enhanced theater missile defense capability later this decade; a national missile deployment readiness program as a hedge against the emergence of a long-range ballistic missile threat to the U.S.

Intelligence Sector Guidelines. The Director, Central Intelligence (DCI) is the lead agency. Guidelines include:

- The US will conduct satellite photo-reconnaissance for peaceful purposes, including intelligence collection and monitoring arms control agreements.

- Ensure that the intelligence space sector provides timely information and data to support foreign, defense and economic policies; military operations; diplomatic activities; indications and warning; crisis management; and treaty verification.

- The existence of the National Reconnaissance Office (NRO) and identification and official titles of its senior officials is unclassified.

Bush Administration Space Policy

Bush's policy stated the primary goal of space activity is to ensure the security of the U.S. and that the U.S. will conduct those activities in space that are necessary for national defense. This policy recognized that space is important in achieving national security, scientific, technical, economic, and foreign policy goals.

The Policy recognized that for five decades, the United States led the world in space exploration and use and developed a solid civil, commercial, and national security space foundation. Space activities improved life in the United States and around the world, enhancing security, protecting lives and the environment, speeding information flow, serving as an engine for economic growth, and revolutionizing the way people view their place in the world and the cosmos. Space had become a place that is increasingly used by a host of nations, consortia, businesses, and entrepreneurs.

The conduct of US space programs and activities were to be a top priority, guided by the following principles:

• The United States is committed to the exploration and use of outer space by all nations for peaceful purposes, and for the benefit of all humanity. Consistent with this principle, "peaceful purposes" allow US defense and intelligence-related activities in pursuit of national interests;

• The United States rejects any claims to sovereignty by any nation over outer space or celestial bodies, or any portion thereof, and rejects any limitations on the fundamental right of the United States to operate in and acquire data from space;

• The United States will seek to cooperate with other nations in the peaceful use of outer space to extend the benefits of space, enhance space exploration, and to protect and promote freedom around the world;

• The United States considers space systems to have the rights of passage through and operations in space without interference. Consistent with this principle, the United States will view purposeful interference with its space systems as an infringement on its rights;

• The United States considers space capabilities -- including the ground and space segments and supporting links -- vital to its national interests. Consistent with this policy, the United States will: preserve its rights, capabilities, and freedom of action in space; dissuade or deter others from either impeding those rights or developing capabilities intended to do so; take those actions necessary to protect its space capabilities; respond to interference; and deny, if necessary, adversaries the use of space capabilities hostile to US national interests;

• The United States will oppose the development of new legal regimes or other restrictions that seek to prohibit or limit US access to or use of space. Proposed arms control agreements or restrictions must not impair the rights of the United States to conduct research, development, testing, and operations or other activities in space for US national interests; and

• The United States is committed to encouraging and facilitating a growing and entrepreneurial US commercial space sector. Toward that end, the United States Government will use US commercial space capabilities to the maximum practical extent, consistent with national security.

The fundamental goals of this policy were to:

• Strengthen the nation's space leadership and ensure that space capabilities are available in time to further US national security, homeland security, and foreign policy objectives;

- Enable unhindered US operations in and through space to defend our interests there;

- Implement and sustain an innovative human and robotic exploration program with the objective of extending human presence across the solar system;

- Increase the benefits of civil exploration, scientific discovery, and environmental activities;

- Enable a dynamic, globally competitive domestic commercial space sector in order to promote innovation, strengthen US leadership, and protect national, homeland, and economic security;

- Enable a robust science and technology base supporting national security, homeland security, and civil space activities; and

- Encourage international cooperation with foreign nations and/or consortia on space activities that are of mutual benefit and that further the peaceful exploration and use of space, as well as to advance national security, homeland security, and foreign policy objectives.

In order to achieve the goals of this policy, the United States Government shall:

- Develop Space Professionals.

- Improve Space System Development and Procurement.

- Increase and Strengthen Interagency Partnerships.

- Strengthen and Maintain the US Space-Related Science, Technology, and Industrial Base.

Obama Administration Space Policy

The current policy published on June 28, 2010 is available online. It outlines a more cooperative tone for international relations.

Leading Collaborative, Responsible, and Constructive Use of Space

The space age began as a race for security and prestige between two superpowers. The decades that followed have seen a radical transformation in the way we live our daily lives, in large part due to our use of space. The growth and evolution of the global economy have ushered in an ever-increasing number of nations and organizations using space to observe and study our Earth, create new markets and new technologies, support operational responses to natural disasters, enable global communications and international finance, enhance security, and expand our frontiers. The impacts of our utilization of space systems are ubiquitous, and contribute to increased transparency and stability among nations.

In a world where the benefits of space permeate almost every facet of our lives, irresponsible acts in space can have damaging consequences for all of us. As such, all nations have a responsibility to act to preserve the right of all future generations to use and explore space. The United States is committed to addressing the challenges of responsible behavior in space, and commits further to a pledge of cooperation, in the belief that with strengthened international cooperation and reinvigorated U.S.

leadership, all nations will find their horizons broadened, their knowledge enhanced, and their lives greatly improved.

Key Elements of the Administration's National Space Policy

- The United States remains committed to many long-standing tenets in space activities. The United States recognizes the rights of all nations to access, use, and explore space for peaceful purposes, and for the benefit of all humanity.
- The United States calls on all nations to share its commitment to act responsibly in space to help prevent mishaps, misperceptions, and mistrust. The United States will take steps to improve public awareness of government space activities and enable others to share in the benefits of space through conduct that emphasizes openness and transparency.
- The United States will engage in expanded international cooperation in space activities. The United States will pursue cooperative activities to the greatest extent practicable in areas including: space science and exploration; Earth observations, climate change research, and the sharing of environmental data; disaster mitigation and relief; and space surveillance for debris monitoring and awareness.
- The United States is committed to a robust and competitive industrial base. In support of its critical domestic aerospace industry, the U.S. government will use commercial space products and services in fulfilling governmental needs, invest in new and advanced technologies and concepts, and use a broad array of partnerships with industry to promote innovation. The U.S. government will actively promote the purchase and use of U.S. commercial space goods and services within international cooperative agreements.
- The United States recognizes the need for stability in the space environment. The United States will pursue bilateral and multilateral transparency and confidence building measures to encourage responsible actions in space, and will consider proposals and concepts for arms control measures if they are equitable, effectively verifiable, and enhance the national security of the United States and its allies. In addition, the United States will enhance its space situational awareness capabilities and will cooperate with foreign nations and industry to augment our shared awareness in space.
- The United States will advance a bold new approach to space exploration. The National Aeronautics and Space Administration will engage in a program of human and robotic exploration of the solar system, develop new and transformative technologies for more affordable human exploration beyond the Earth, seek partnerships with the private sector to enable commercial spaceflight capabilities for the transport of crew and cargo to and from the International Space Station, and begin human missions to new destinations by 2025.
- The United States remains committed to the use of space systems in support of its national and homeland security. The United States will invest in space situational awareness capabilities and launch vehicle technologies; develop the means to assure mission essential functions enabled by space; enhance our ability to identify and characterize threats; and deter, defend, and if necessary, defeat efforts to interfere with or attack U.S. or allied space systems.
- The United States will fully utilize space systems, and the information and applications derived from those systems, to study, monitor, and support responses to global climate change and natural disasters. The United States will accelerate the development of satellites to observe and study the Earth's environment, and conduct research programs to study the Earth's lands, oceans, and atmosphere.

DoD Space Policy

Consistent with the National Space Policy, Department of Defense space forces will continue to support military operations worldwide, monitor and respond to strategic military threats, and monitor arms control and nonproliferation agreements and activities. DoD will exploit and, if required, control space to assist in the successful execution of the National Security Strategy and National Military Strategy.

DoD Space Policy focuses on operational capabilities that enable the military services to fulfill national security objectives. The policy supports and amplifies US national space policy. Space is recognized as being a medium within which the conduct of military operations in support of our national security can take place, just as on land, at sea, and in the atmosphere, and similarly from which military space functions of space support, force enhancement, space control and force application can be performed. The DoD has not updated the Space Policy pursuant to issuance of the National Space Policy of June 2010; however the principles support the current policy.

Key features of the current policy are:

• Space is a medium like the land, sea, and air within which military activities shall be conducted to achieve US national security objectives. The ability to access and utilize space is a vital national interest because many of the activities conducted in the medium are critical to US national security and economic well being.

• Ensuring the freedom of space and protecting US national security interests in the medium are priorities for space and space-related activities. US space systems are national property afforded the right of passage through and operations in space without interference.

• The primary DoD goal for space and space-related activities is to provide operational space force capabilities to ensure that the United States has the space power to achieve its national security objectives. Contributing goals include sustaining a robust US space industry and a strong, forward-looking technology base. (This section goes on to further describe how pace activities shall contribute to the achievement of US national security objectives.)

• Mission Areas. Capabilities necessary to conduct the space support, force enhancement, space control, and force application mission areas shall be assured and integrated into an operational space force structure that is sufficiently robust, ready, secure, survivable, resilient, and interoperable to meet the needs of the POTUS and SECDEF, Combatant Commanders, Military Services, and intelligence users across the conflict spectrum.

• Assured Mission Support. The availability of critical space capabilities necessary for executing national security missions shall be assured. Access to space, robust satellite control, effective surveillance of space, timely constellation replenishment and reconstitution, space system protection, and related information assurance, access to critical electromagnetic frequencies, critical asset protection, critical infrastructure protection, force protection, and continuity of operations shall be ensured to satisfy the needs of the POTUS AND SECDEF, Combatant Commanders, Military Services, and the intelligence users across the conflict spectrum.

- Planning. Planning for space and space-related activities shall focus on improving the conduct of national security space operations, assuring mission support, and enhancing support to military operations and other national security objectives. Such planning shall also identify missions, functions. and tasks that could be performed more efficiently and effectively by space forces than terrestrial alternatives.

The DoD Space Policy defines space missions as Space Support, Force Enhancement, Space Control, and Force Application.

- Space Support functions are those required to deploy and maintain military equipment and personnel in space. They include activities such as launching and deploying satellites, maintaining and sustaining space vehicles while in orbit (Telemetry, tracking, and commanding), and recovering space vehicles, if required. It includes providing logistics support for the space, ground control and launch elements. It provides for surge launch capabilities to replace loss of space assets and the acquisition of replacement satellites.

- Force Enhancement are defined as any operation conducted from space with the objective of enhancing, enabling, or supporting terrestrial operations in peacetime, conflict and war. Force enhancement includes such capabilities as communications, position and navigation, weather and remote sensing, reconnaissance, intelligence, surveillance, missile launch detection and warning support at the tactical, operational, and strategic levels of war. Also, civil/commercial/allied capabilities may augment DoD systems to support military space force enhancement requirements, particularly if primary DoD capabilities were to be lost. The efficiencies resulting from the use of these space capabilities can have a dramatic effect on Army operations: reducing uncertainty, facilitating battle command, and moderating the effects of friction and fog of war.

- Space Control consists of operations that ensure freedom of action in space for friendly forces while limiting or denying enemy freedom of action. It includes the conduct of offensive and defensive space operations to prevent an enemy's space forces from gaining and maintaining space superiority and to ensure survivability and protection of friendly space systems. These missions include naval, air, land, and space operations that disrupt, deny and destroy an adversary's capability to use space systems. It also includes satellite negation and as well as space and ground segment protection. The Army's role in this function will be from the terrestrial perspective, such as jamming up/downlink frequencies and attacking satellite control nodes and facilities from the ground.

- Force application is conducted primarily from or through space with the intent to destroy surface and subsurface targets for the purpose of ballistic missile defense (BMD) or power projection. While force application capabilities from space are limited, the role of space in force application is evolving. It consists of the offensive and defensive use of space and space-related capabilities to project combat power and defend US military forces and their allies from attack. In the broadest sense, any space system capable of providing and disseminating information contributes to force application. Consistent with treaty obligations and national policy, this capability could include the use of space- and ground-based systems to provide protection from ballistic missiles, in programs such as national missile defense (NMD) and theater missile defense (TMD), and to extend the Army's force projection range against surface targets.

ARMY SPACE POLICY OVERVIEW

The U.S. Army is one of the largest users of space-based capabilities in DOD. As the Army transforms, its operational characteristics will, in large part, be achieved through the use and exploitation of transformational space systems. This dependency requires the Army to actively participate in defining space related capability needs that ensure necessary force structure and systems are developed and acquired to enable the land force to conduct the full range of military operations now and in the future.

The Army's four broad space related objectives are:

- To maximize the effectiveness of current space capabilities in support of operational and tactical land warfighting needs.

- To influence the design, development, acquisition, and concepts of operation of future space systems that enable and enhance current and future land forces.

- To advance the development and effective use of responsive, timely, and assured Joint interoperable space capabilities.

- To seamlessly integrate relevant space capabilities into the operating force.

DOCTRINE

Joint Pub 3-14

The latest Joint Pub 3-14, *Joint Doctrine for Space Operations,* was published in January 2009. The publication sets forth doctrine, and selected tactics, techniques, and procedures to govern the joint activities and performance of the Armed Forces of the United State in joint operations as well as the doctrinal basis for U.S. military involvement in multinational and interagency operations. It defines the military operational principles associated with supporting from space and operating in space. The command relationships and responsibilities need updating in accordance with the Unified Command Plan changes since the date of publication.

Key concepts in Joint Pub 3-14 are:

- Joint Forces have come to rely heavily on space capabilities to provide them unprecedented "battlespace awareness". Space, air, land and sea commanders must be aware of the characteristics operational considerations and constraints inherent to the space region in order to maximize capabilities and forces to support joint operations.

- Commander, USSTRATCOM can be either a supported or supporting commander.

- Commander, USSTRATCOM has planning and operational responsibility for the region of space, and exploits this fourth operating medium to bring unique capabilities to the joint battlespace. CDR, USSTRATCOM maintains COCOM of DoD space forces. Space assets and space capabilities can be made OPCON and/or TACON to another Unified Command/JFC/JTF Commander.

<u>FM 3-14</u>

The latest FM 3-14, *Space Support to Army Operations*, published in January 2010, focuses on the use of space capabilities across the full range of military operations. In this regard, FM 3-14 is more than a doctrinal statement about space support to the Army. It provides space support doctrine that is not only consistent with current doctrine of the various mission areas, but should drive the future development of doctrine within those areas. The bottom line is to aggressively use space to support the attainment of terrestrial objectives. It is relevant from the highest levels of command down to the soldier in the foxhole.

<u>Additional Relevant Space Doctrine</u>

While this text will not specifically discuss all space service doctrine, it is worth noting USAF space doctrine. Two documents conceptualize how the USAF envisions space operations impacting and integrating as part of our military operations. Air Force Doctrine Document (AFDD) 2-2, *Space Operations*, is dated 27 Nov 2006 and AFDD 2-2.1, *Counterspace Operations*, is dated 2 Aug 2004. Both documents complement related discussion found in Joint Pub 3-14 and in FM 3-14.

<u>Concept Capability Plan (CCP) Space Operations</u> (TRADOC Pam 525-7-4, 15 November 2006)

As part of implementing the Army Space Policy, the Army has established TRADOC Pamphlet 525-7-4, Concept Capability Plan for Space Operations 2015-2024.

The US Army Space Operations Concept Capability Plan (CCP) identifies the space-enabled capabilities required to execute Army operations in the 2015 - 2024 timeframe. The CCP describes how Army forces leverage the power of national, civil and commercial space-based assets on the future battlefield. The capabilities identified in this CCP provide a coherent way ahead for the further examination of potential doctrine, organization, training, materiel, leadership and education, personnel and facilities (DOTMLPF) solutions. As such, this CCP will serve as a start point for a comprehensive capabilities based assessment involving many different proponents.

The CCP is designed to achieve four imperatives:

- Facilitate the integration of space capabilities across the full spectrum of Army and joint operations.

- Improve the Army's ability to exploit existing space capabilities.

- Deliver space capabilities that address Army needs (capability requirements) and priorities by influencing the design of space-based systems and payloads.

- Systematically and deliberately evolve Army space support operations over time to provide dedicated, responsive theater focused support to operational and tactical commanders.

PLAN FOR OPERATIONALLY RESPONSIVE SPACE

This report was developed to meet requirements outlined in the National Defense Authorization Act for Fiscal Year 2007 (P.L. 109-364), which provided that the Secretary of Defense submit to the congressional defense committees a report setting forth a plan for the acquisition by the Department of

Defense of capabilities for operationally responsive space to support military users and military operations.

Operationally Responsive Space (ORS) has been defined broadly in DoD as assured space power focused on timely satisfaction of Joint Force Commanders' needs. This definition considers ORS as a subset of space activities designed to satisfy Joint Force Commanders' (JFCs') needs, while also maintaining the ability to address other users' needs, for improving the responsiveness of space capabilities to meet national security requirements.

The ORS initiative will create opportunities for integration and operational efficiencies needed to ensure affordable access to the space-based capabilities that are critical to fulfilling the full range of US diplomatic, information, military, and economic needs - a specific goal of NSPD-49. New approaches to methods, development, and acquisition are necessary to attain ORS capabilities and the broader space operations efficiency.

Table 1-1

International Agreements that Limit

Military Activities in Space

Agreement	Principle/Constraint
United Nations Charter (1947)	Made applicable to space by the Outer Space Treaty of 1967.
	Prohibits states from threatening to use, or actually using, force against the territorial integrity or political independence of another state (Article 2(4)).
	Recognizes a state's inherent right to act in individual or collective self-defense when attacked. Customary international law recognizes a broader right to self-defense, one that does not require a state to wait until it is actually attacked before responding. This right to act preemptively is known as the right of anticipatory self-defense (Article 51).
Limited Test Ban Treaty (1963)	Bans nuclear weapons tests in the atmosphere, in outer space, and underwater.
	States may not conduct nuclear weapon tests or other nuclear explosions (i.e., peaceful nuclear explosions) in outer space or assist or encourage others to conduct such tests or explosions Article I).
Outer Space Treaty (1967)	Outer space, including the Moon and other celestial bodies, is free for use by all states (Article I).
	Outer space and celestial bodies are not subject to national appropriation by claim of sovereignty, use, occupation, or other means (Article II).
	Space activities shall be conducted in accordance with international law, including the UN Charter (Article III).
	The Moon and other celestial bodies are to be used exclusively for peaceful purposes (Article IV).
	Nuclear weapons and other weapons of mass destruction (such as chemical and biological weapons) may not be placed in orbit, installed on celestial bodies, or stationed in space in any other manner (Article IV).
	A state may not conduct military maneuvers; establish military bases, fortifications or installations; or test any type of weapon on celestial

bodies.   Use of military personnel for scientific research or other peaceful purpose is permitted (Article IV).

Table 1-1 Continued

Agreement	Principle/Constraint
Outer Space Treaty (1967)	States are responsible for governmental and private space activities, and must supervise and regulate private activities (Article IV).
	States are internationally liable for damage to another state (and its citizens) caused by its space objects (including privately owned ones) (Article VII).
	States retain jurisdiction and control over space objects while they are in space or on celestial bodies (Article VII).
	States must conduct international consultations before proceeding with activities that would cause potentially harmful interference with activities of other parties (Article IX).
	States must carry out their use and exploration of space in such a way as to avoid harmful contamination of outer space, the Moon, and other celestial bodies, as well as to avoid the introduction of extraterrestrial matter that could adversely affect the environment of the Earth (Article IX).
	Stations, installations, equipment, and space vehicles on the Moon and other celestial bodies are open to inspection by other countries on a basis of reciprocity (Article XII).
Agreement on the Rescue Astronauts (1968)	The Agreement, elaborating on elements of articles 5 and 8 of the of Outer Space Treaty, provides that States shall take all possible steps to rescue and assist astronauts in distress and promptly return them to the launching State, and that States shall, upon request, provide assistance to launching States in recovering space objects that return to Earth outside the territory of the Launching State.
Antiballistic Missile (ABM) Treaty (1972)*	Originally between the US and USSR, then between the US and Russia. (Now no longer in effect.)
	Prohibited development, testing, or deployment of space-based ABM systems or components (Article V).
	Prohibited deployment of ABM systems or components except as authorized in the treaty (Article I).

Prohibited interference with the national technical means a party uses to verify compliance with the treaty (Article XII).

As of December 2001 the US, President Bush, announced to Russia that the US was pulling out of the ABM Treaty.

Table 1-1 Continued

Agreement	Principle/Constraint
Liability Convention (1972)	A launching site is absolutely liable for damage by its space object to people or property on the Earth or in its atmosphere (Article II).
	Liability for damage caused by a space object, to persons or property on board such a space object, is determined by fault (Article III).
Convention on Registration (1976)	Requires a party to maintain a registry of objects it launches into Earth orbit or beyond (Article II).
	Information of each registered object must be furnished to the UN as soon as practical, including basic orbital parameters and general function of the object (Article IV).
Environmental Modification Convention (1980)	Prohibits military or other hostile use of environmental modification techniques as a means of destruction, damage, or injury to any other state if such use has widespread, long-lasting, or severe effects (Article I).
Agreement Governing the Activities of States on the Moon and Other Celestial Bodies (1984)	The Agreement reaffirms and elaborates on many of the provisions of the Outer Space Treaty as applied to the Moon and other celestial bodies, providing that those bodies should be used exclusively for peaceful purposes, their environments should not be disrupted, that the United Nations should be informed of the location and purpose of any station established on those bodies. In addition, the Agreement provides that the Moon and its natural resources are the common heritage of mankind and that an international regime should be established to govern th exploitation of such resources when such exploitation is about to become feasible.

Chapter 2
Organizations

OVERVIEW

There are several organizations responsible for DoD space operations. United States Strategic Command (USSTRATCOM), responsible for the command and control of US strategic forces, assumed responsibility for DoD space and information operations from US Space Command on October 1, 2002. USSTRATCOM provides space support for unified commanders worldwide. This chapter addresses the responsibilities of various space organizations.

UNITED STATES STRATEGIC COMMAND (USSTRATCOM)

 US Strategic Command is one of nine US unified commands under the Department of Defense. Headquartered at Offutt Air Force Base, Neb., USSTRATCOM is a global integrator charged with the missions of full-spectrum global strike, space operations, computer network operations, Department of Defense information operations, strategic warning, integrated missile defense, global C4ISR (Command, Control, Communications, Computers, Intelligence, Surveillance, and Reconnaissance), combating weapons of mass destruction, and specialized expertise to the joint warfighter.

Day-to-day planning and execution for the primary mission areas is done by five Joint Functional Component Commands or JFCCs and three other functional components:

• U.S. Cyber Command -- Plans, coordinates, integrates, synchronizes, and conducts activities to: direct the operations and defense of specified Department of Defense information networks and; prepare to, and when directed, conduct full-spectrum military cyberspace operations in order to enable actions in all domains, ensure US/Allied freedom of action in cyberspace and deny the same to our adversaries.

• JFCC-Integrated Missile Defense (IMD) -- develops desired characteristics and capabilities for global missile defense operations and support for missile defense. Plans, integrates and coordinates global missile defense operations and support (sea, land, air and space-based) for missile defense.

• JFCC-Intelligence, Surveillance and Reconnaissance (ISR) -- plans, integrates and coordinates intelligence, surveillance and reconnaissance in support of strategic and global operations and strategic deterrence. Tasks and coordinates ISR capabilities in support of global strike, missile defense and associated planning.

• **JFCC-Space (JFCC SPACE)** -- Continuously coordinates, plans, integrates, commands and controls space operations to provide tailored, responsive, local and global effects, and on order, denies the enemy the same, in support of national, USSTRATCOM, and combatant commander objectives.

• JFCC-Global Strike (JFCC GS) -- Conducts kinetic (nuclear and conventional) and non-kinetic effects planning. GS manages global force activities to assure allies and to deter and dissuade actions detrimental to the United States and its global interests; should deterrence fail, employs global strike forces in support of combatant commander.

- Joint Information Operations Warfare Command (JIOWC) -- plans, integrates, and synchronizes Information Operations (IO) in direct support of Joint Force Commanders and serves as the USSTRATCOM lead for enhancing IO across DoD.

- USSTRATCOM Center for Combating Weapons of Mass Destruction (SCC-WMD) -- The USSTRATCOM Center for Combating WMD (SCC-WMD) is collocated with the Defense Threat Reduction Agency at Fort Belvoir, Virginia. The director of DTRA leads the new center which integrates and synchronizes Department of Defense-wide efforts in support of the combating WMD mission. The SCC-WMD serves to plan, advocate and advise the commander, U.S. STRATCOM on WMD-related matters. The center provides recommendations to dissuade, deter and prevent the acquisition, development or use of WMD and associated technology. Through collaboration with US and allied organizations, the SCC-WMD leverages around-the-clock situational awareness of worldwide WMD and related activities, as well as provides day-to-day and operational crisis support via the operations center. DTRA, with its extensive worldwide information sharing and expertise in the chemical, biological, radiological and nuclear fields, provides critical reachback and resources to the SCC-WMD and USSTRATCOM, and other combatant commands.

USSTRATCOM provides space support to deployed US military forces worldwide and defends the Defense Information Infrastructure, the computer and communications networks, of the Department of Defense against unauthorized intrusion or attack. The command's space missions consist of:

- Space Support: Launching and operating satellites. Includes satellite operations and telemetry, tracking, and commanding (TT&C), and spare activation. All launches occur at Cape Canaveral Air Force Station, Fla., or Vandenberg Air Force Base, Calif.

- Force Enhancement: Satellite communications, navigation, weather, missile warning and intelligence.

- Space Control: Assuring US access to and freedom of operation in space, and denying enemies the same.

- Force Application: Researching and developing space-based capabilities that have the potential to engage adversaries from space. Requires policy change before implementation.

USSTRATCOM Space Components:
- Air Force Space Command, Peterson AFB, Colo.

- US Army Space and Missile Defense Command, Redstone Arsenal, AL.

- Fleet Forces Command, Little Creek Naval Air Station, Va.

AIR FORCE SPACE COMMAND (AFSPC)

AFSPC was established on 1 September 1982 to consolidate Air Force space activities and is responsible for operating assigned military space systems. Located at Peterson AFB, Colorado, it is a major command of the US Air Force and is the Air Force service component which supports the USSTRATCOM mission through its functional component, the 14th Air Force. AFSPC is also a component of USSTRATCOM for Inter-Continental Ballistic Missile forces; 20th Air Force provides the ICBM forces for AFSPC.

AFSPC's mission is to provide resilient and cost-effective Space and Cyberspace capabilities for the Joint Force and the Nation.

Organization

Fourteenth Air Force is located at Vandenberg AFB, Calif., and provides space capabilities for the joint fight through the operational missions of spacelift; position, navigation and timing; satellite communications; missile warning and space control.

Twenty-fourth Air Force is located at Lackland AFB, Texas, and its mission is to provide combatant commanders with trained and ready cyber forces which plan and conduct cyberspace operations. The command extends, operates, maintains and defends its assigned portions of the Department of Defense network to provide capabilities in, through and from cyberspace.

The Space and Missile Systems Center at Los Angeles AFB, Calif., designs and acquires all Air Force and most Department of Defense space systems. It oversees launches, completes on-orbit checkouts and then turns systems over to user agencies. It supports the Program Executive Office for Space on the Global Positioning, Defense Satellite Communications and MILSTAR systems. SMC also supports the Evolved Expendable Launch Vehicle, Defense Meteorological Satellite and Defense Support programs and the Space-Based Infrared System.

The Space Innovation and Development Center at Schriever AFB, Colo., is responsible for integrating space systems into the operational Air Force. The mission is to advance full-spectrum warfare through rapid innovation, integration, training, testing and experimentation.

The Air Force Network Integration Center at Scott AFB, Ill., is the Air Force focal point for shaping, providing, sustaining and integrating the Enterprise Network and enabling assured core cyber capabilities to achieve warfighting advantage. AFNIC's vision is to integrate the evolving Enterprise Network environment at the speed of need to achieve information dominance and peerless cyberspace warfighting capabilities.

The Air Force Spectrum Management Office, located in Alexandria, Va., is responsible for planning, providing and preserving access to the electromagnetic spectrum for the Air Force and selected DoD activities in support of national policy objectives, systems development and global operations. AFSMO defends and articulates Air Force spectrum access to regulatory agencies at the joint, national and international levels. It is responsible for all Air Force spectrum management-related matters, policy and procedures. Additionally, the agency oversees the Air Force spectrum management career field and manages the payment of the approximately $4 million Air Force spectrum fee each year.

AFSPC bases include: Schriever, Peterson and Buckley, Colo.; Los Angeles and Vandenberg, Calif., and Patrick, Fla. In addition, many geographically separated units span the globe.

Spacelift operations at the East and West Coast launch bases provide services, facilities and range safety control for the conduct of DOD, NASA and commercial launches. Through the command and control of all DOD satellites, satellite operators provide force-multiplying effects -- continuous global coverage, low vulnerability and autonomous operations. Satellites provide essential in-theater secure communications, weather and navigational data for ground, air and fleet operations and threat warning. Ground-based radar and Defense Support Program satellites monitor ballistic missile launches around the world to guard against a surprise missile attack on North America. Space surveillance radars provide vital information on the location of satellites and space debris for the nation and the world. Maintaining space superiority is an emerging capability required to protect our space assets. With a readiness rate above 99 percent, America's ICBM team plays a critical role in maintaining world peace and ensuring the nation's safety and security.

AFSPC acquires, operates and supports the Global Positioning System, Defense Satellite Communications Systems, Defense Meteorological Satellite Program, Defense Support Program and the Space-Based Infrared System Program. AFSPC currently operates the Delta II, Delta IV and Atlas V launch vehicles. The Atlas V and Delta IV launch vehicles comprise the Evolved Expendable Launch Vehicle program, which is the future of assured access to space. AFSPC's launch operations include the Eastern and Western ranges and range support for all launches, including the space shuttle on the Eastern Range. The command maintains and operates a worldwide network of satellite tracking stations, called the Air Force Satellite Control Network, to provide communications links to satellites.

Ground-based radars used primarily for ballistic missile warning include the Ballistic Missile Early Warning System, PAVE Phased Array Warning System and Perimeter Acquisition Radar Attack radars. The Maui Optical Tracking Identification Facility, Ground-based Electro-Optical Deep Space Surveillance System, Passive Space Surveillance System, phased-array and mechanical radars provide primary space surveillance coverage.

ARMY SPACE AND MISSILE DEFENSE COMMAND/ARMY STRATEGIC COMMAND

The U.S. Army Space and Missile Defense Command/Army Strategic Command (SMDC/ARSTRAT) provides command and control to the 1st Space Brigade and the 100th Missile Defense Brigade (Ground-based Midcourse Defense). SMDC also provides secure, space-based Blue Force Tracking and communications planning through Regional Satellite Communications (SATCOM) Support Centers and the Spectral Operations Resource Center to Army forces and, upon request, to joint warfighters.

In its role as the the Army service component to USSTRATCOM, the command is called Army Forces Strategic Command – ARSTRAT. The SMDC also serves as the Army's specified proponent for Space and National Missile Defense and as the Army's integrator for Theater Missile Defense.

The command ensures that Army warfighters have access to space assets and the products they provide to win decisively with minimum casualties - and effective missile defense to protect our nation as well as our deployed forces and those of our friends and allies.

From its headquarters in Huntsville, AL, the U.S. Army SMDC oversees a number of Army elements around the globe to accomplish its challenging and diverse mission.

1st Space Brigade in Colorado Springs, CO, is the Army's first and only space brigade. Elements of the brigade's three battalions deployed to Afghanistan and later to Iraq and the surrounding theater. 1st Space Brigade conducts continuous, global space support, space control and space force enhancement operations in support of the STRATCOM and supported combatant commanders enabling the delivery of decisive combat power.

53rd Signal Battalion (SATCON) is responsible for the daily command and control of the Defense Satellite Communications System (DSCS) and the Wideband Global SATCOM System (WGS) communications networks supported by these satellites. DSCS provides reliable, robust, worldwide, continuous communications support to U.S. warfighting forces, strategic military users, the U.S. intelligence community and numerous national agencies. The battalion's mission is to manage, plan, and control the payloads of the DSCS and WGS satellites. The battalion operates and maintains five companies that operate control facilities located around the world. At Schreiver Air Force Base in Colorado Springs, the DSCS Certification Facility (DCF) forms the Headquarters Company for the battalion. The 53rd Signal Battalion (SATCON) controls the satellite links for tactical and strategic warfighter communications networks. The companies also provide payload control to the satellite as well as technical and troubleshooting assistance required to ensure maximum support to the user. In addition, the DCF provides platform control, monitoring the health and welfare of the payloads for selected satellites in the DSCS and WGS constellation. On a typical day, the Wideband SATCOM Operations centers control nearly 1,000 links providing vital communications support to deployed warfighters, strategic users, and the intelligence community around the world.

1st Space Battalion was activated on 15 Dec 99. The battalion mission is to plan, coordinate, and integrate Space Force Enhancement functions and Space Control Operations in support of Army, joint, and combined forces. The battlion consists of HHC, the 1st Space Company (theater missile warning), 2nd Space Company (space support), and the 4th Space Company (space integrated ground suite).

117th Space Battalion is a Colorado National Guard battalion. Soldiers in the 117th Space Battalion enable operations by maximizing military utilization of space-based assets to include sattelite imagery, missile warning systems, sattelite communications, space-based weather and global positioning system capabilities.

The subordinate companies that fall under the battalion are the 217th Space Company and the 1158th Space Company; both located near Peterson AFB in Colorado Springs.

100th Missile Defense Brigade (Ground-based Midcourse Defense), Colorado Army National Guard, provides oversight of the Soldiers trained to operate the nation's limited missile defense capability. The brigade comes under the overall direction of the responsible combatant commander during an operational mission.

49th Missile Defense Battalion, Alaska Army National Guard, provides physical security and defense of the interceptor site as well as operators who are trained to fire the missiles.

NAVAL NETWORK WARFARE COMMAND

Naval Network Warfare Command (NETWARCOM) is the Navy's central operational authority for space, information technology requirements, network and information operations in support of naval forces afloat and ashore; to operate a secure and interoperable naval network that will enable effects-based operations and innovation; to coordinate and assess the Navy operational requirements for and use of network/command and control/information technology/information operations and space; to serve as the operational forces' advocate in the development and fielding of information technology, information operations and space and to perform such other functions and tasks as may be directed by higher authority.

Naval Network Warfare Command serves as a Functional Component Commander to U.S. Strategic Command (STRATCOM).

Several commands will be under the authority of the commander of NETWARCOM: Naval Network and Space Operations Command (NNSOC) in Dahlgren, Va.; Fleet Information Warfare Center (FIWC) in Norfolk, Va.; and Navy Component Task Force Computer Network Defense (NCTF CND) in Washington, D.C. NNSOC will be established coincident with NETWARCOM, through the merger of the existing Naval Space Command and the Naval Network Operations Command. The Space and Naval Warfare Systems Command (SPAWAR) will report for additional duties to the NETWARCOM commander for matters related to fleet support and execution-year requirements. The commander of Naval Security Group (NAVSECGRU) will serve additional duties as the NETWARCOM director of Information Operations.

NAVAL SATELLITE OPERATIONS CENTER (NAVSOC)

NAVSOC, headquartered in Point Mugu, California, originally established as the Navy Astronautics Group in 1962 to operate the Navy Navigation Satellite System, commonly known as TRANSIT. The Command operates remote satellite ground control stations in Prospect Harbor, Maine; Laguna Peak, California; and Finegayan, Guam. In addition, NAVSOC uses a state-of-the-art distributed mission control system to maintain telemetry, tracking, and control (TT&C) on operational and scientific satellites, such as: FLTSATCOM satellites, UHF Follow-on (UFO) including GBS, Geodetic Satellite Follow-on (GFO), and Navy Ionospheric Monitoring System (NIMS) satellite constellation. NAVSOC also provides on-orbit technical and engineering support in conjunction with spacecraft operations.

NATIONAL RECONNAISSANCE OFFICE

The NRO is a joint organization engaged in the research and development, acquisition, launch and operation of overhead reconnaissance systems necessary to meet the needs of the Intelligence Community and of the Department of Defense. The NRO conducts other activities as directed by the Secretary of Defense and/or the Director of National Intelligence. A DoD agency, the NRO is staffed by DoD and CIA personnel. It is funded through the National Reconnaissance Program, part of the National Foreign Intelligence Program.

The NRO designs, builds and operates the nation's reconnaissance satellites. NRO products, provided to an expanding list of customers like the Central Intelligence Agency (CIA) and the Department of Defense (DoD), can warn of potential trouble spots around the world, help plan military operations, and monitor the environment. As part of the 16-member Intelligence Community, the NRO plays a primary role in achieving information superiority for the U. S. Government and Armed Forces.

THE NATIONAL GEOSPATIAL-INTELLIGENCE AGENCY

 NGA provides timely, relevant and accurate geospatial intelligence in support of national security objectives. The term "geospatial intelligence" means the exploitation and analysis of imagery and geospatial information to describe, assess and visually depict physical features and geographically referenced activities on the Earth. Geospatial intelligence consists of imagery, imagery intelligence and geospatial (e.g., mapping, charting and geodesy) information.

NGA is a member of the US Intelligence Community and a Department of Defense (DoD) Combat Support Agency. Headquartered in Bethesda, Md., NGA operates major facilities in the St. Louis, Mo. and Washington, D.C. areas. The Agency also fields support teams worldwide.

Information collected and processed by NGA is tailored for customer-specific solutions. By giving customers ready access to geospatial intelligence, NGA provides support to civilian and military leaders and contributes to the state of readiness of US military forces. NGA also contributes to humanitarian efforts, such as tracking floods and disaster support, and to peacekeeping. NGA products include:

- geospatial intelligence in all its forms, and from whatever source—imagery, imagery intelligence and geospatial data and information—to ensure the knowledge foundation for planning, decision and action

- tailored, customer-specific geospatial intelligence, analytic services and solutions

The Director of NGA is also the functional manager for the National System for Geospatial Intelligence (NSG). The NSG integrates the technology, policies, capabilities and doctrine necessary to conduct geospatial intelligence in a multi-intelligence environment.

The NSG is the combination of technology, policies, capabilities, doctrine, activities, people and communities necessary to produce geospatial intelligence in an integrated multi-intelligence, multi-domain environments.

The membership of the NSG includes the Intelligence Community, the Joint Staff, the Military Departments (to include the Services), the Combatant Commands (COCOMs), international partners, National Applications Office, Civil Applications Committee members, industry, academia, Defense service providers, and civil community service providers.

NATIONAL SECURITY AGENCY

 The NSA mission is to protect US communications and produce foreign intelligence information. Supply leadership, products, and services to protect classified and unclassified information from interception, unauthorized access, and technical intelligence threats. In the foreign signals intelligence area, the central point for collecting and processing activities

conducted by the US government, with authority to produce signals intelligence in accord with objectives, requirements, and priorities established by the CIA director with the advice of the National Foreign Intelligence Board.

It was established by a Presidential directive in 1952 as a separate agency within DoD under the direction, authority, and control of the Secretary of Defense, who serves as the executive agent of the US government for the production of communications intelligence information. The Central Security Service was established in 1972 by a Presidential memorandum to provide a more unified cryptological organization within the Defense Department. The NSA director also serves as chief of the CSS and controls the signals intelligence activities of the military services.

Chapter 3
The Space Environment

OVERVIEW

Increased dependence on space-based systems to meet warfighter objectives and needs, coupled with the increasing use of microelectronics and a move to non-military specifications for satellites, increases vulnerability to loss of critical satellite functions or entire systems. Warfighters must understand the space environment in order to know the impact and potential effect on satellites and ground systems.

Space is often incorrectly thought of as a vast, empty vacuum that begins at the outer reaches of the Earth's atmosphere and extends throughout the universe. In reality, space is a dynamic place that is filled with energetic particles, radiation, and trillions of objects both very large and very small. Compared to what we experience on Earth, it is a place of extremes. Distances are vast. Velocities can range from zero to the speed of light. Temperatures on the sunny side of an object can be very high, yet extremely low on the dark side, just a short distance away. Charged particles continually bombard exposed surfaces. Some have so much energy that they pass completely through an object in space. Magnetic fields can be intense. The environment in space is constantly changing, and is a hostile environment for satellites.

Where Space Begins

There is no formal definition of where space begins. International law, based on a review of current treaties, conventions, agreements and tradition, describes the lower boundary of space as the lowest perigee attainable by an orbiting space vehicle. A specific altitude is not mentioned. By international law standards aircraft, missiles and rockets flying over a country are considered to be in its national airspace, regardless of altitude. Orbiting spacecraft are considered to be in space, regardless of altitude. The US government defines space in the same terms as international law.

The Earth's atmosphere does not suddenly end at a particular altitude and space begins. In fact, the Earth's atmosphere continues out for more than 1,000 miles into space. In practical terms, the lowest altitude for a satellite in a circular orbit is about 93 miles (150 km) but, without propulsion, the satellite would quickly lose speed and fall back to Earth.

Space Weather

Our space weather originates at the sun. The area between the Sun and the planets has been termed the interplanetary medium. Fifty years ago, most scientists believed that Earth was surrounded by an empty, unchanging vacuum. The launches of the first satellites beginning in 1957 (the former USS.R.'s Sputnik was the first) changed all that. We now know that space is filled with debris from disintegrated comets, an ever-changing high-speed solar wind, radiation belts and polar fountains. Bad weather in space can mean poor radio communications, dilution of GPS accuracy, or disrupted power grids here on Earth, and potential hazards to astronauts or spacecraft can be even more serious. Space is actually a turbulent area dominated by the solar wind, which flows at velocities of approximately 250-1000 km/s (about 600,000 to 2,000,000 miles per hour). It is filled with low energy charged particles, photons, electric and magnetic fields, dust, and cosmic rays. The densities of these things are low, particularly for the particulate matter, compared to, for example, the atmospheric density at the surface of the Earth but they are high enough to affect spacecraft, humans in space and even occasionally human activities on Earth. Moreover, other characteristics of the solar wind (density, composition, and magnetic field strength, among others) vary with changing conditions on the Sun that can cause violent and dramatic changes in our atmosphere. These may be visibly manifested in dramatic aurora displays. This is called space weather. Table 3-1 shows how space weather is analogous to common Earth weather measurements and features:

Earth Weather	Comparable Space Weather
Wind speed	Solar wind speed
Wind direction	Particle direction
Isobars	Magnetic flux lines
Weather systems	Magnetic field areas (ex: van Allen belts)

Table 3-1: Comparison of Earth Weather to Space Weather

The solar wind flows around obstacles such as planets, but the magnetic fields of those planets respond in specific ways. Earth's magnetic field is very similar to the pattern formed when iron filings align around a bar magnet. Under the influence of the solar wind, these magnetic field lines are compressed in the sunward direction and stretched out in the leeward direction. This creates the magnetosphere (Figure 3-1a), a complex, teardrop-shaped cavity around Earth.

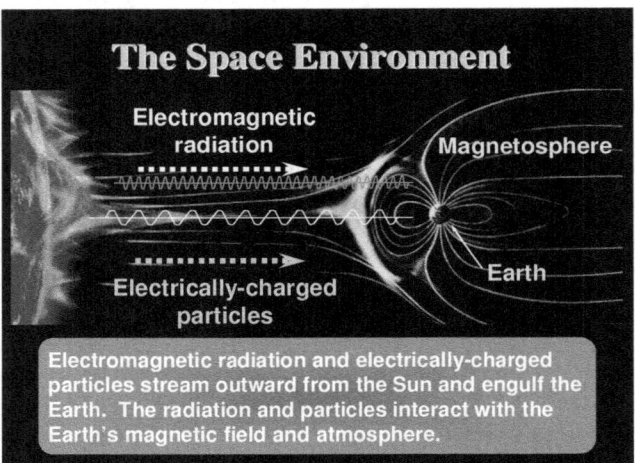

Figure 3-1a: The Space Environment

The Van Allen radiation belts are within this cavity, as is the ionosphere, a layer of Earth's upper atmosphere where photo ionization by solar x-rays and extreme ultraviolet rays creates free electrons. Earth's magnetic field reacts to the solar wind its speed, density, and magnetic field. Because the solar wind varies over time scales as short as seconds, the interface that separates interplanetary space from the magnetosphere is very dynamic. Normally this interface called the magnetopause lies at a distance equivalent to about 10 Earth radii in the direction of the Sun. However, during episodes of elevated solar wind density or velocity, the magnetopause can be pushed inward to within 6.6 Earth radii (the altitude of geosynchronous satellites). As the magnetosphere extracts energy from the solar wind, internal processes produce geomagnetic storms.

To understand the effects of space weather, one must understand the properties of the various layers of the earth's atmosphere, the properties of the sun and solar energy and the properties of the earth's electromagnetic field. We will begin by discussing the layers of the earth's atmosphere and the properties of those layers as they relate to satellite orbits.

EARTH'S ATMOSPHERE

The atmosphere limits the lowest altitude at which a satellite can be placed into orbit. The atmosphere absorbs, diffuses, deflects, or delays certain frequencies of signals sent to and from a satellite. Satellites launched from the Earth's surface must pass through the atmosphere to attain orbit. Manned spacecraft and some unmanned payloads must reenter the atmosphere to safely return to the surface. The Earth's atmosphere is divided into numerous regions which have different characteristics (Figure 3-1b). The boundaries between the regions are not distinct. Some regions overlap and others are made up of a number of sub-regions. Definitions are complicated by the fact that different scientific fields define these regions in different ways using various criteria, such as pressure or temperature. Altitude alone does not define where one region ends and another begins because the regions are constantly fluctuating in size depending on the time of day, the season of the year, the degree of activity of the sun and many other factors.

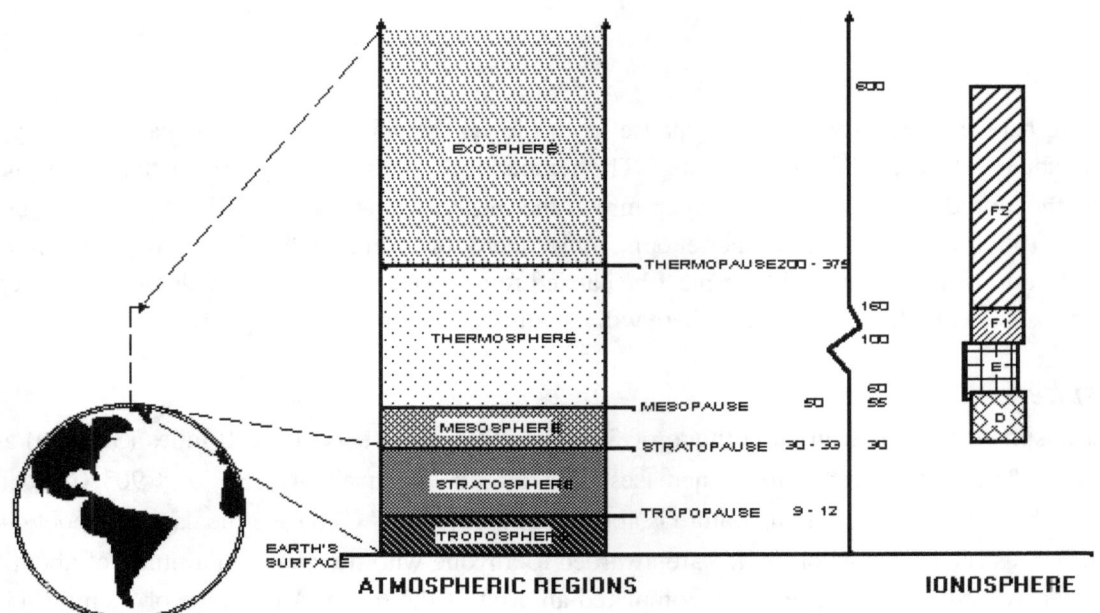

Figure 3-1b: Earth's Atmosphere

Troposphere

The troposphere is the lowest region of the atmosphere. It starts at the Earth's surface and extends to the tropopause, the upper boundary of the troposphere. The altitude of the tropopause varies from 9 to 12 miles (15 to 20 km) at the equator to about 6 miles (10 km) in polar areas. Almost all clouds and weather occur in the troposphere. It contains about 99% of the atmosphere's water vapor and 90% of the air. Above an altitude of 2 miles (3.2 km), a person who has not become adapted requires supplemental oxygen or a pressurized environment to survive. Approximately half of the Earth's atmosphere is below an altitude of 3 miles (5 km).

Stratosphere

The layer above the troposphere is the stratosphere. It extends from the tropopause to the stratopause, the upper boundary at about 30 to 33 miles (48 to 53 km) altitude. In general, the temperature of the stratosphere increases slightly with altitude which results in vertical stability. Approximately 99 percent of the atmosphere is in the stratosphere and troposphere. This region is characterized by the near absence of water vapor and clouds. At altitudes above 9 miles (14.5 km) breathing supplemental oxygen through a mask is no longer effective. Pressurization by means of a pressure cabin or a pressure suit and helmet is required. Above an altitude of 15 miles (24 km) compressing outside air into the cabin or pressure suit usually generates too much heat. Everything required to sustain life must be carried on board. Ozone is in the ozone layer which varies in altitude from 12 to 21 miles above the Earth. Ozone is poisonous, therefore, in the stratosphere the outside atmosphere cannot be used to pressurize a crew cabin. The ozone layer is important because it absorbs a large portion of the sun's ultraviolet radiation which is harmful to humans and most other life forms. In Figure 3-1b, the small band around the picture of the Earth represents the combined thickness of the troposphere and stratosphere relative to the size of the Earth.

Mesosphere

The mesosphere extends from the stratopause at the lower boundary to the mesopause, the upper boundary at about 50 miles (80 km) altitude. The temperature of the atmosphere in the mesosphere decreases as the altitude increases. The temperature at the mesopause is about -130° F (-90° C). Above about 30 miles (48 km) altitude there is not enough atmosphere for even a high altitude ramjet to operate. Above this altitude both fuel and oxidizer must be carried for a rocket engine to provide thrust. Meteors and space debris entering the earth's atmosphere will usually burn up in the mesosphere.

Thermosphere

The thermosphere extends from an altitude of 50 miles (80 km) to between 200 miles (320 km) and 375 miles (600 km). The temperature increases with altitude from about -130° F (-90° C) to the thermopause where the maximum temperature is about 2,960° F (1,475° C) during the day and about 440° F (225° C). US astronaut wings or device are awarded to anyone who travels to an altitude of about 50 miles or higher, regardless of whether they completed an orbit of the Earth. An altitude of 93 miles (150 km) is the lowest altitude at which a satellite in a circular orbit can orbit the Earth for at least one revolution without propulsion. At this altitude it takes 87.5 minutes to complete one revolution of Earth. This altitude is the most commonly accepted definition of where space begins but it is not explicitly stated in any treaty or international agreement. An altitude of 80 miles (129 km) is about the lowest altitude (perigee) at which a satellite in an elliptical orbit can pass through the Earth's atmosphere and still remain in orbit. This is discussed further in Chapter 4.

Exosphere

The exosphere begins where the thermosphere ends and extends out into space. In this region the density of atoms and molecules of the atmospheric gases is very low. Typically, individual atoms travel about 1600 miles in 20 minutes before colliding with another atom. After the collision, some atoms have enough velocity to escape the Earth's gravity and enter interplanetary space. The density is so low that all of the atmospheric particles that surround the Earth at an altitude of 1,000 miles could be contained in 1 cm3 at sea level. Satellites orbiting in the exosphere at an altitude of about 620 miles (1000 km) are, however, slowed by atmospheric drag caused by friction from collisions with individual molecules.

Ionosphere

The ionosphere is a region that is defined by its high density of ionized molecules, rather than by temperature changes. It begins within the mesosphere, between 31 and 50 miles in altitude and extends to about 240 miles (400 km). The atoms and molecules in this layer are bombarded by solar X-rays and ultraviolet radiation from the sun and become electrically charged, or ionized, in a process known as photoionization. This process produces an excess of free electrons and ionized atoms. X-rays and extreme ultraviolet (EUV) radiation from the sun and ultra-high-frequency galactic cosmic rays from the stars of outer space are the prime mechanisms in the formation of the ionosphere. The amount of radiation available to ionize atmospheric molecules varies between day and night, high and low latitude, and with the activity level of the sun. Sunspots, solar flares and other disturbances on the surface of the Sun produce fluctuations in the output of the Sun's energetic rays and particles. These fluctuations cause Sudden Ionospheric Disturbances (SIDs). Simply put, when particles interact with magnetic fields they give off energy which is seen as lights. Interaction between the ionosphere and the concentrated magnetic field of the Earth at the North and South Poles causes the undulating brilliant colors of the aurora borealis (north lights) and the aurora australis (southern lights).

The ionosphere can absorb, delay or reflect certain frequencies of radio signals; therefore, the characteristics of the ionosphere have significant impact on the design and operation of communications systems between satellites and ground stations. The ionosphere is also important for ground-to-ground radio communications systems because they rely on the ability of the ionosphere to reflect or refract radio waves to achieve long range.

Magnetosphere

The magnetosphere is that area of space, around the Earth, that is controlled by the Earth's magnetic field. The solar wind compresses the Earth's magnetic field on the side closest to the Sun creating the magneto pause. Therefore, this area changes in relation to changes in the Sun's activity. The magnetosphere impedes the direct entry of the solar wind plasma into the earth's atmosphere. This area and the resulting effects on satellites will be discussed in further detail.

THE SUN

The origin of space environmental impacts on radar, communications and space systems lies primarily with the sun. The sun continuously emits electromagnetic energy and charged particles in the form of light and radio frequency noise. Superimposed on these emissions are enhancements in the electromagnetic radiation (particularly at X-ray, Extreme Ultra Violet (EUV) and Radio wavelengths) and in the energetic charged particle streams emitted by the sun. The constant stream of these forms of radiation is called the solar wind. In addition, at seemingly irregular intervals, there are solar flares, the explosive ejection of particles (mostly protons and electrons) accompanied by sporadic emissions of electromagnetic radiation.

The Solar Wind

The sun flings 1 million tons of matter out into space every second. Because of the high temperature of the Sun's corona, solar protons and electrons acquire velocities in excess of the escape velocity from the sun. The result is that there is a continuous outward flow of charged particles in all directions from the sun. This flow of particles is called the solar wind. And this happens every day, day after day, year after year.

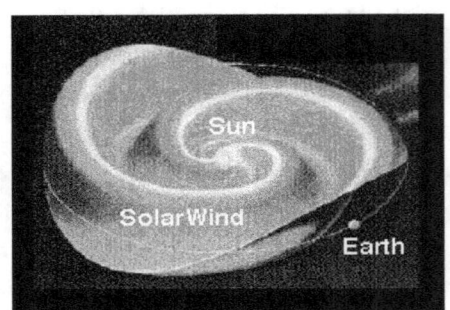

Figure 3-2a: Solar Wind

The solar wind is a fully ionized gas that literally explodes continuously from the solar corona, the outer region of the sun's dense atmosphere carrying with it magnetic fields still attached to the sun. The solar wind is driven away from the sun by thermal pressure from the two million-degree corona at supersonic velocities, about 500 km/s. By the time the solar wind reaches Earth's orbit, it is traveling at 185 - 435 mi/sec (well over 1,000,000 miles per hour).

The density is 1 - 10 particles per cubic centimeter. When it gets as far away as the orbit of the earth, 150 million km, it is still accelerating and it keeps on going as far as we have been able to probe with our most distant spacecraft.

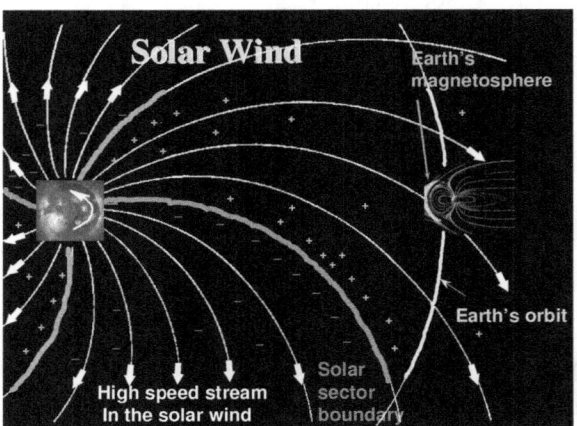

Figure 3-2b: Solar Wind

The Interplanetary Magnetic Field (IMF) emanating from the sun normally has 4 to 6 sectors of alternating positive and negative polarity, and a spiral structure near the plane of the Earth's orbit due to the sun's 27-day rotation period. The IMF guides charged solar wind particles, and those in one IMF sector can't penetrate into another sector. Since particles tend to move faster in a sector's forward rather than tailing portion, particle density tends to build up at "Solar Sector Boundaries (SSBs)." Also dense

"High Speed Streams (HSSs)" of particles can exist within a sector. There are regions in the sun's atmosphere where magnetic field lines are open to space and actually facilitate the outward flow of particles. Both SSB and HSS enhancements and discontinuities can disrupt the Earth's magnetosphere as the sun's rotation causes them to sweep by the Earth.

These solar wind enhancements and discontinuities occur throughout the solar cycle, even during Solar Minimum. Fortunately, the size of the disruptions they cause in the Earth's magnetosphere tends to be less than with the disruptions caused by solar flares. Also, since these solar wind enhancements and discontinuities are tied to solar features that persist for longer than the sun's 28-day rotation period, they tend to be recurrent, and thus the geomagnetic storms they produce are somewhat easier to forecast than the sporadic geomagnetic storms produced by spontaneous flares that can occur at any time.

The velocity and density of the solar wind vary with sunspot activity. Gusts and disturbances form in the solar wind associated with violent events on the sun. The gusting and blowing of the solar wind against the Earth's protective magnetic shield in space is responsible for space weather storms. Some of these disturbances are described in the following sections.

Sun Spots

The Sun has a rotation period of 27 days, which exposes Earth to the surface features of the Sun, such as sunspots. One way we track solar activity is by observing sunspots. Sunspots are relatively cool areas that appear as dark blemishes on the face of the sun. They are formed when magnetic field lines just below the sun's surface are twisted and poke though the solar photosphere. The twisted magnetic fields above sunspots are sites where solar flares are observed to occur, and we are now beginning to understand the connection between solar flares and sunspots. Sunspots are normally associated in a complex, but not completely understood, way with solar flares, i.e., the more the sunspots, the more the solar flares

Every 11 years the sun undergoes a period of activity called the "solar maximum", followed by a period of quiet called the "solar minimum." During the solar maximum there are many sunspots, solar flares, and coronal mass ejections, all of which can affect communications and weather here on Earth. During solar minimum there are few events. The last two solar maximums occurred around 1989 and then again at the beginning of the year 2001. The next is predicted to occur on or about May 2013.

Solar Flares

A solar flare is an explosion of incredible power and violence that releases energy equivalent to about 100 hurricanes in a matter of tens of minutes. (To equal the power of a hurricane one would have to set off about a thousand nuclear devices per second for as long as the hurricane rages on.) It involves sudden bursts of particle acceleration, plasma heating, and bulk mass motion. The strong and twisted magnetic fields in the vicinity of active sunspot groups are thought to provide the power that is

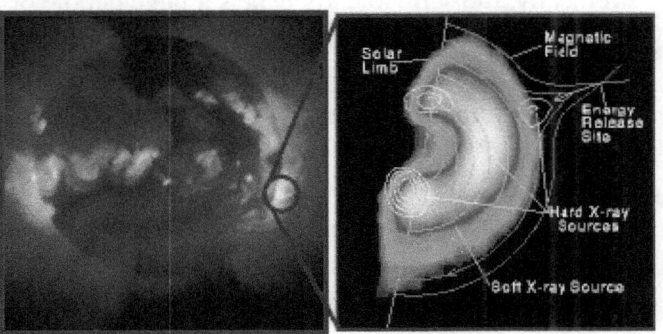

Figure 3-3: Parts of a Solar Flare

released in the solar flares. It is not known as yet exactly how this occurs.

In the largest flares, 1032 ergs or more can be released in a few minutes to a few tens of minutes. Such large flares only occur a few times within a year or two of the maximum in solar activity that occurs every 11 years or so. Many smaller flares occur down to the limits of detectability of modern instruments at about 1027 ergs. These smaller events generally last for shorter times down to a few seconds; their occurrence rate also follows the 11-year cycle, peaking at several tens of flares per day.

In general, a solar flare produces copious radiation across the full electromagnetic spectrum from the longest wavelength radio waves to the highest energy gamma rays. Flares release energy in many forms - electro-magnetic (Gamma rays and X-rays), energetic particles (protons and electrons), and mass flows. The contrast over the background (quiet-Sun) emission is much higher at the shorter X-ray and gamma ray wavelengths. The X-rays result from the interactions of the high energy electrons energized during the flare, and the gamma rays result primarily from nuclear interactions of the high energy protons and other heavier ions. The X-rays travel at the speed of light and arrive at Earth in just 8 minutes time. The hot particles, which travel at slower speeds, follow several hours later.

There are many events that may occur on Earth following a solar flare. By the time we first observe a flare, it is already causing immediate environmental effects and satellite system impacts. These impacts are almost entirely limited to the Earth's sunlit hemisphere, as the radiation does not penetrate or bend around the earth. In addition to the increase in visible light, minutes later there is a Sudden Ionosphere Disturbance (SID) in Earth's ionosphere. This, in turn, causes short wave fade-out, resulting in the loss of long-range over-the-horizon communications for 15 minutes to 1 hour. During the first few minutes of a flare, there may be a radio noise storm. The first few minutes of this storm causes noise over a wide range of frequencies that can be heard as static in radio transmissions. These effects will be discussed in more detail. The National Oceanic and Atmospheric Administration (NOAA) monitors the X-Ray flux from the Sun with detectors on some of its satellites. Observations for the last few days are available at NOAA's website for Today's Space Weather.

Coronal Mass Ejection

Perhaps the most important solar event for the people on Earth is a coronal mass ejection (CME) because it can produce spectacular space weather events at Earth. Periodically, the sun violently expels a huge bubble of gas (called a coronal mass ejection) from its outer atmosphere into space. During the course of such an event on average, several billion tons of gases are blown toward Earth. This is equivalent to the mass contained in a small lake but vaporized and traveling at millions of miles an hour (thousands of kilometers per second). The total energy contained in these high-speed charged gases during a large event is roughly the same as a large solar flare

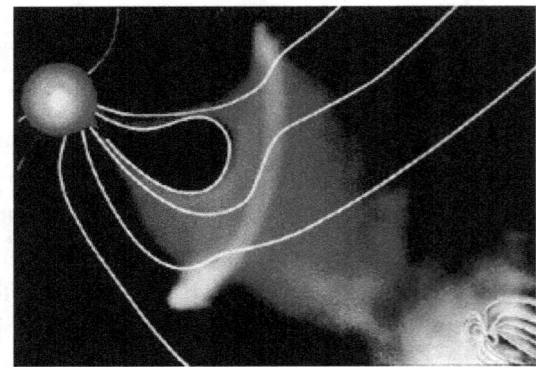

Figure 3-4: Coronal Mass Ejection with Shock Waves

(approximately equivalent to the total energy of 100 hurricanes) but expelled over a longer time period than a solar flare (many hours as compared to tens of minutes).

CMEs disrupt the flow of the solar wind and produce disturbances that strike the Earth with sometimes catastrophic results. If a CME collides with the Earth, it can excite a geomagnetic storm. Large geomagnetic storms have, among other things, caused electrical power outages and damaged satellites. In space CMEs typically drive shock waves that produce energetic particles that can be damaging to both electronic equipment and astronauts that venture outside the protection of the Earth's magnetic field. Solar flares, on the other hand, directly affect the ionosphere and radio communications at the Earth, and also release energetic particles into space. Therefore, to understand and predict "space weather" and the effect of solar activity on the Earth, an understanding of both CMEs and flares is required.

This high speed plasma bubble expands rapidly into space rivaling the sun in size in just a matter of hours. CMEs normally drive shocks ahead of them as they plow into the slower-moving solar wind streams (Figure 3-4). Vast quantities of high speed particles are born in these shock structures and speed ahead of the coronal mass ejection to impact the Earth before the CME itself makes an appearance. It is now believed that a large fraction of the solar energetic particles arriving at Earth were produced in these shocks rather than at flare sites on the Sun.

CMEs are difficult to detect and, in fact, were not observed until coronagraphs were built and flown in space. The frequency of CMEs varies with the sunspot cycle. At solar minimum we observe about one CME a week. Near solar maximum we observe an average of 2 to 3 CMEs per day. Many of the CMEs that are blown toward the Earth never actually directly hit us. They pass us by.

An international group of scientist discovered a new way of predicting coronal mass ejections. Using the Japanese Yohkoh satellite, it was discovered that an S-shaped structure often appears on the Sun in advance of the violent eruptions. Early warning of approaching solar storms is very useful to power companies, the communications industry and any space operations. NASA now gains information from the satellites STEREO (Solar TEerrestrial RElations Observatory), ACE (Advanced Composition Explorer) and SOHO (SOlar and Heliospheric Observatory).

Other Solar Activities

One type of corpuscular radiation is the cosmic ray. Cosmic rays originate from two sources: the Sun (solar cosmic rays), and other stars throughout the universe (galactic cosmic rays). This radiation is primarily high velocity protons and electrons. The solar cosmic rays are not a serious threat to humans, except during periods of solar flare activity, when the radiation can increase a thousand fold over short periods. Cosmic ray particles can also cause direct damage to internal components through collision. Shielding is not feasible due to the high energy of the particles and the weight of the shielding required. Cosmic rays have the most impact on polar and geosynchronous orbits. This is due to the fact that they are outside or near the edge of the protective shielding provided by Earth's magnetic field.

Thermal energy emitted by the Sun is intense. Satellites in orbit around the Earth do not have the benefit of the shielding provided by the atmosphere. Parts of satellites exposed to sunlight can heat to very high temperatures if they absorb the radiant energy. Sunlight reflected by the Earth is also a significant source of radiant energy for satellites in low Earth orbit. On the shadow side, heat can radiate from the satellite into cold space with the result that surface temperatures can be hundreds of degrees below zero. Thermal energy is also generated by internal components of a satellite such as batteries, transmitters, computers and other devices. This thermal energy must be dissipated so that it does not damage the components.

A Hypothetical Timeline of a Solar Event

Within 8 Minutes....

X-ray and ultraviolet light given off by solar flares arrives at the Earth. These types of light are responsible for the formation of the ionosphere under normal circumstances. But now the intensities of these types of light have increased dramatically. All over the dayside hemisphere, the ionosphere is increased in density, particularly at low altitudes in the D and E regions (see Figure 3-1b). This is called a sudden ionospheric disturbance (SID) and lasts usually for between 10 and 60 minutes. Short wave radio signals experience increased absorption when ionization is produced at these low altitudes. Loss of signal at this time is called a short wave fadeout. Enhanced levels of solar radiation can also cause heating and expansion of the neutral atmosphere and increase the amount of atmospheric drag that a satellite experiences in an unpredictable manner.

Minutes to Hours Later...

Solar protons are accelerated to very high velocities in the explosive release of energy associated with a solar flare. These particles first make their way out of the corona. Then they cannot take a straight line path to the Earth but must travel along the Sun's extended magnetic field lines (called the interplanetary magnetic field or IMF). These field lines have a structure resembling the spiral traced out by a spinning lawn sprinkler, which is referred to as a Parker spiral.

Hours to Days Later...

Disturbances in the solar wind arrive at the Earth within hours to days after a violent event on the Sun. The largest space weather disturbances at Earth are produced by coronal mass ejections and fast solar wind streams emanating from coronal holes. Coronal holes are the main source of recurrent solar activity.

ELECTROMAGNETIC FORCES

Earth's Magnetic Field

The Earth has a magnetic field which emanates from its south magnetic pole, extends out into space and comes back to its north magnetic pole. The north and south magnetic poles are near the north and south celestial poles about which the Earth rotates. The magnetic field around the Earth is similar to one that would be formed if a bar magnet extended through the center of the Earth with the tips at the north and south magnetic poles.

According to accepted theory, the Earth's magnetic field is thought to be produced by electric currents circulating in the

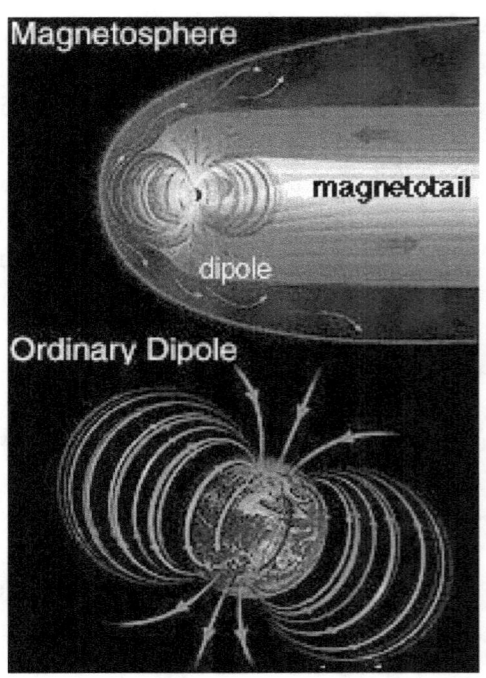

Figure 3-5: Dipole vs. Magnetosphere Magnetic Fields

molten core; however, the exact mechanism is not yet understood. This magnetic field extends outward from the Earth's core into interplanetary space where it encounters the magnetic field and moving charged plasma of the solar wind. The solar wind flows around the Earth's magnetic field but distorts the field as it does so. The solar wind compresses the Earth's magnetic field on the dayside and stretches it out into a long anti-sunward tail on the night side.

Magnetosphere

The solar wind hits the earth's magnetic field. Because the solar wind is mostly charged particles, electrons and protons, it cannot easily penetrate the earth's field. It meets an immovable obstacle and it forms a shock front and tries to flow around the sides in a manner very much analogous to a supersonic aircraft moving through the atmosphere. In so doing, it compresses and confines the magnetic field on the side toward the sun and stretches it out into a long tail on the night side. The magnetic cavity formed by this process is called the "magnetosphere" and is shown in Figure 3-5. In the process, energy and charged particles are transferred from the solar wind to the magnetosphere.

The space around our atmosphere is alive and dynamic because the Earth's magnetic field reacts to changes in the solar wind. The interaction between the solar wind and the plasma of the magnetosphere acts like an electric generator, creating electric fields deep inside the magnetosphere. These fields in turn give rise to a general circulation of the plasma within the magnetosphere and accelerate some electrons and ions to higher energies. During periods of highest solar wind, powerful magnetic storms in space near the Earth cause vivid auroras, radio and television static, power blackouts, navigation problems for ships and airplanes with magnetic compasses, and damage to satellites and spacecraft. Events on the Sun and in the magnetosphere can also trigger changes in the electrical and chemical properties of the atmosphere, the ozone layer, and high-altitude temperatures and wind patterns.

On the sunward side of Earth, the solar wind compresses the magnetic field in toward the Earth, increasing the magnetic field strength in the

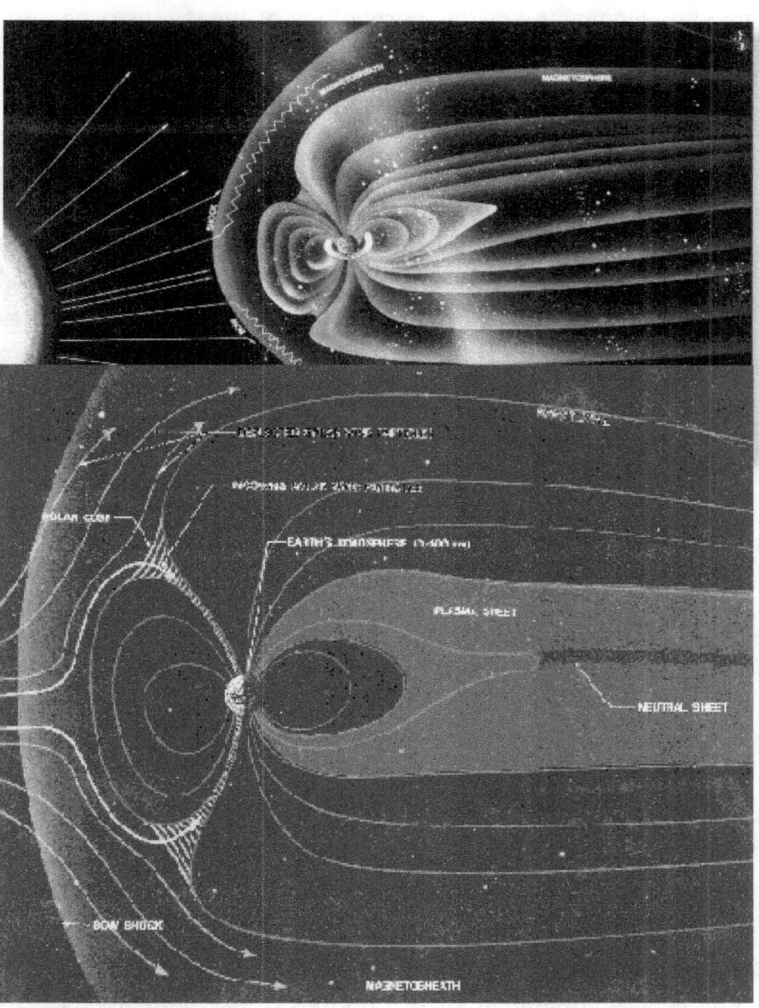

Figure 3-7: Earth's Magnetosphere

compressed areas. On the opposite side of Earth, the solar wind acts to stretch out the magnetic field thus giving it a teardrop shape. The boundary of the magnetosphere is the magnetopause. It is where the pressure of the solar wind is balanced by Earth's magnetic field pressure. The magnetopause is most defined on the sunward side where it is located approximately 10 Earth radii (10 times 3962 mi) from the Earth. This boundary fluctuates between 7 to 14 Earth radii during magnetic disturbances resulting from large variations in the solar wind.

On the side of Earth opposite from the Sun, the solar wind draws the magnetic field out into a long tail, called the magnetotail. This tail extends out to 1,000 earth radii or more. Located within the magnetotail is a region of high density, high energy plasma, that is known as the plasma sheet. It may extend out past 300 earth radii. Within the plasma sheet is the neutral sheet. This is where the magnetic field lines reverse direction from a component towards Earth (northern lobe) to a component away from Earth (southern lobe).

Since the beginning of the space age in 1957, and even before, scientists have been probing and defining regions of the magneto-sphere and trying to understand how they arise. We now have a fairly good understanding of the morphology of the magnetosphere. The magnetic field lines from the dayside of the magnetosphere are drawn over the poles and into the night side to form this long stretched tail. In the process, they form the northern and southern lobes which are separated from each other by a very interesting region called the plasma sheet. In the plasma sheet, by some process not yet fully understood, the magnetic energy of the lobes is converted to plasma kinetic energy. Plasma is a gas that is fully ionized, i.e. the atoms, mostly hydrogen, has had all of their electrons removed. Although almost unknown to us on earth, considering the whole universe, plasma is the most common form of matter, being found mostly in stars. Plasma is generally considered to be electrically neutral, but the fact that the atoms are ionized means that the weak magnetic and electric fields that pervade the region strongly influence the charged particle motion. In fact, the definition of the magnetosphere is that region where the earth's magnetic field controls the particle behavior. The story of the dynamics of the magnetosphere is one of motion of particles from the solar wind and the earth's ionosphere being shoved around and energized by electric and magnetic fields.

Van Allen Radiation Belts

Within the magnetosphere are found the Van Allen radiation belts, named after Dr. James Van Allen, who was responsible for the interpretation of the data from the Explorer I satellite, launched by the US Army. Their existence was not discovered until early 1958, when Explorer I data was used to map particles trapped by Earth's magnetic field. Data from this satellite and other satellites revealed the strange shape of the magnetosphere and toroidal shaped pockets of trapped charged particles. Other scientific satellites have added even more detail. The Van Allen radiation belts consist of concentric doughnut shaped regions of energetically charged particles (Figure 3-8). The Van Allen belts

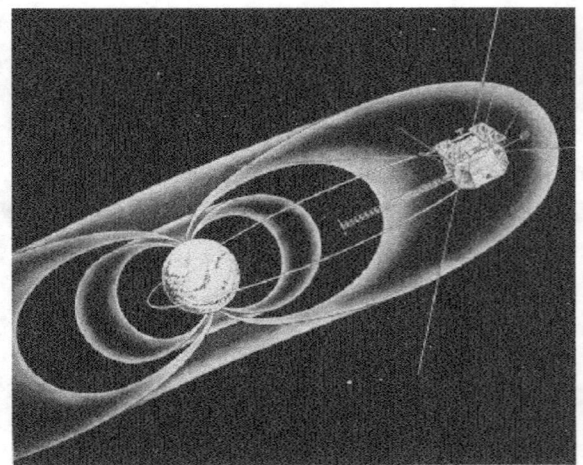

Figure 3-8: Van Allen Radiation Belts

have an inner and outer portion filled with high energy protons and electrons. The protons are most intense at about 2,200 miles. The electron flux breaks at about 10,000 miles. The area of low particle density separating the two belts is often called the "Slot." Although the cross sectional view shown in Figure 3-8 shows two distinct belts, they are not really that distinct. The density of electrons in the slot is just somewhat lower than in the two Van Allen radiation belts.

The inner Van Allen radiation belt starts at an altitude between 250 miles to 750 miles, depending upon the latitude. It extends to about 6,200 miles. This is where the "slot" begins. The inner belt extends from about 40° north latitude to about 40° south latitude.

The outer Van Allen belt begins at about 6,200 miles (depending on latitude) and extends to between 37,000 and 57,000 miles. The upper boundary is dependent upon the activity of the sun. When electrons, protons, and perhaps some other charged particles encounter Earth's magnetic field, many of them are trapped by the field. They bounce back and forth between the magnetic north and south poles, following the magnetic field lines.

Experience has shown that spacecraft (manned or unmanned) in low circular orbits (120-340 mi) receive an insignificant amount of radiation from the Van Allen zones. A satellite in a geosynchronous orbit, however, could be close enough to the center of the outer zone (22,500 mi) to accumulate a hazardous dose. For manned space systems, the spacecraft must be shielded and an orbit that minimizes radiation exposure must be selected. Other satellites in other-than-geosynchronous orbits may transition in and out of the Van Allen zones repeatedly, receiving varying amounts of radiation.

Electromagnetic Forces

As a satellite orbits the Earth, it travels through the magnetic fields that cause the satellite to act like a magnet. The electrical or electronic components within or outside the satellite set up magnetic fields, which react with the Earth's magnetic field. Another reason a satellite may act as a magnet is due to a negative electrical charge that is generated by the satellite passing through the partly ionized medium which produces a negative charge on the satellite's skin. The negative charge is higher on the dayside of the orbit than on the night side. The motion of a charged satellite made of conductive materials through Earth's magnetic field also results in the satellite acquiring an electrical potential gradient which is proportional to the intensity of the magnetic field and the velocity of the satellite as it passes through the field. These cause a magnetic drag to act upon the satellite. The drag can cause torquing of the satellite.

AURORAS

The aurora borealis (northern lights) and the aurora australis (southern lights) are multicolored bands of visible light effects seen in the nighttime sky. Auroras occur in the upper atmosphere of both the north and south poles where the Earth's magnetic field bends towards the Earth's surface. The bottom of the aurora curtain is about 62 miles (100 km) in altitude, extending up to about 190 miles (300 km). Seen from above by research satellites, the aurora curtain appears along oval belts around the Earth's

Figure 3-9: Auroras

geomagnetic poles. The average radius is about 1,400 miles from the poles (about 70° north or 70° south latitude) and extends thousands of miles to the east and west. Typically, the aurora band is only about one mile or less thick.

The light of the auroras is caused by electrical discharges powered by the interaction of the Earth's magnetic field and the solar wind. The solar wind supplies the necessary charged particles which perturb the magnetosphere so that the particles are "dumped" into the atmosphere near Earth's magnetic poles. As these particles collide with gas molecules in the upper atmosphere, light energy is released, causing aurora displays. Increased activity in the auroras indicates major solar activity that can affect military communications.

GEOGRAPHIC ANOMALIES

The center of Earth's magnetic field is offset from the center of Earth by 270 miles towards the Western Pacific Ocean. This leads to two areas on Earth where the magnetic field strength will be either stronger or weaker than expected. These areas are geomagnetic anomalies. The South Atlantic anomaly is a region of unusually low magnetic field strength. At this location where the low strength field lines dip to a low altitude, many particles enter the atmosphere and collide with the higher density atmospheric particles. As a result of these collisions, we have ionospheric effects similar to those produced in the aurora regions. This will disrupt high frequency (HF) transmissions. The Southeast Asian anomaly is an area of unusually high magnetic field strengths. As a result of this strong field, the trapped particles will have a higher density at any given altitude. This leads to an enhanced F2 region in the ionosphere. This can adversely affect communication transmissions that pass through this region. The solutions to the problems caused by these anomalies are to use a frequency not affected by the disturbed conditions or to avoid transmissions within either region.

EFFECTS ON SPACE AND GROUND SYSTEMS

Generally the stronger a solar flare, the denser/faster/more energetic a particle stream, or the sharper a solar wind discontinuity or enhancement, the more severe will be the event's impacts on the near-Earth environment and on DoD systems operating in that environment. Unfortunately the satellite system impacts do not occur one at a time, but will most likely occur in combinations of more than one thing. The stronger the causative solar-geophysical activity, the more in number of simultaneous effects a system may experience.

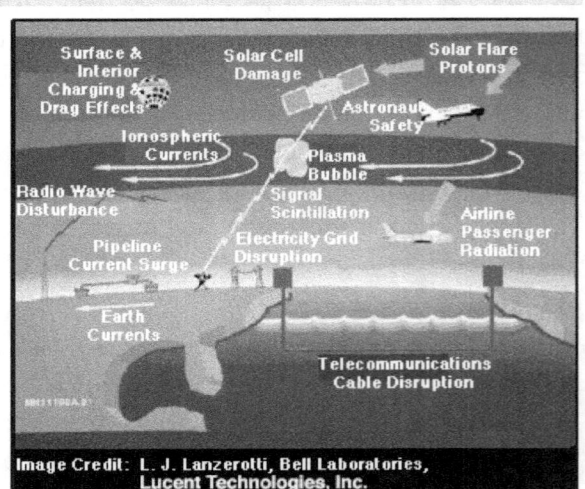

Figure 3-10: Effects of Space Weather on Ground Systems

Ground systems are also affected by solar-geophysical activity (Figure 3-10). Some of these ground system impacts can indirectly affect military operations. For example, system impacts from a geomagnetic storm can include: (1) induced electrical currents in power lines and in long pipelines (such as the Alaskan oil pipeline), which can cause transformer failures and power outages and (2), magnetic field variations, which can lead to compass errors and interfere with geological surveys. Each of the three general categories of solar radiation (Figure 3-11) has its own characteristics and types of system impacts.

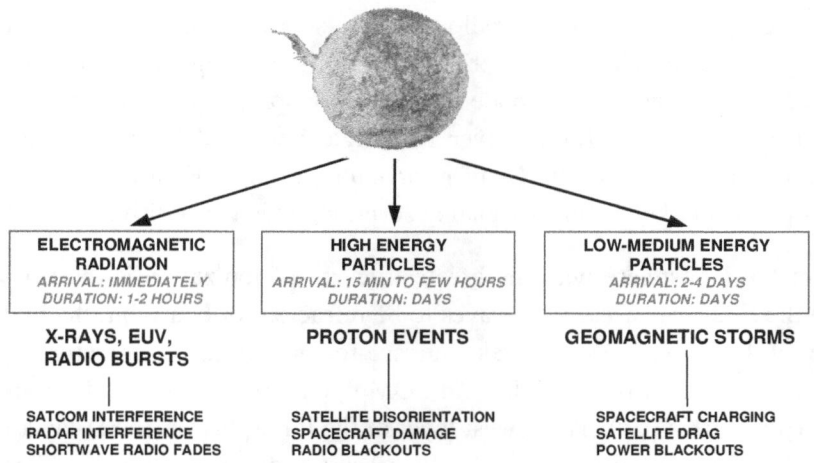

Figure 3-11: Three Categories of Solar Radiation

Every solar event is unique in its exact nature and the enhanced emissions it produces. Some solar events cause little or no impact on the near-Earth environment because their enhanced particle and/or electromagnetic (X-ray, EUV and/or Radio wave) emissions are too feeble or their particle streams may simply miss hitting the Earth. For those events that do affect the near-Earth environment, effects can be both immediate and delayed.

Electromagnetic Radiation

Electromagnetic radiation impacts are almost entirely limited to the Earth's sunlit hemisphere, as the radiation does not penetrate or bend around the earth. Since enhanced electromagnetic emissions cease when the flare ends, the effects tend to subside as well. As a result, these effects tend to last only a few tens of minutes to an hour or two. Sample system effects include: satellite communications (SATCOM) and radar interference (specifically, enhanced background noise), navigation errors and absorption of HF (3-30 MHz) radio communications.

The first of the specific system impacts to be discussed will be the Short Wave Fade

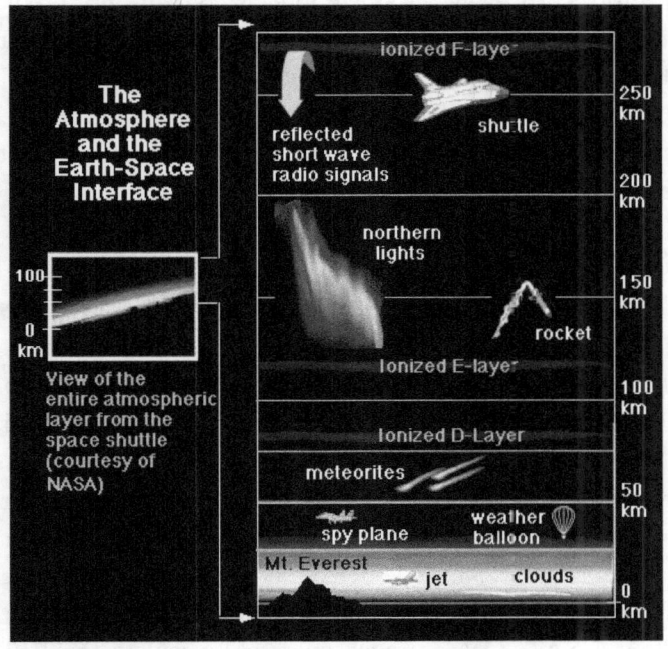

Figure 3-12: Ionospheric Regions

(SWF), which is caused by solar flare X-rays. The second impact covered will be SATCOM and radar interference caused by solar flare radio bursts.

Short Wave Fade (SWF) Events

The High Frequency (HF, 3-30 MHz) radio band is also known as the short wave band. Thus, a SWF refers to an abnormally high fading (or absorption) of a HF radio signal. HF radio waves are refracted by the ionosphere's F-layer. The normal mode of radio wave propagation in the HF range is by refraction using the ionosphere's strongest (or F) layer for single hops and by a combination of reflection and refraction between the ground and the F-layer for multiple hops (Figure 3-12). However, each passage through the ionosphere's D-layer causes signal absorption, which is additive.

The portion of the ionosphere with the greatest degree of ionization is the F-layer. The presence of free electrons in the F-layer causes radio waves to be refracted (or bent), but the higher the frequency, the less the degree of bending. As a result, surface-to-surface radio operators use Medium or High Frequencies (300 KHz to 30 MHz), while SATCOM operators use Very to Extremely High Frequencies (VHF/EHF 30 MHz to 300 GHz). The lowest layer of the ionosphere is the D-layer (normally between 50 and 90 km altitude). The D-layer acts to absorb passing radio wave signals. This has the effect of the lower the frequency then the greater the degree of signal absorption.

X-ray radiation emitted during a solar flare can significantly enhance D-layer ionization and absorption over the entire sunlit hemisphere of the Earth. This enhanced absorption is known as a SWF and may, at times, be strong enough to close the HF propagation window completely (called a Short Wave Blackout). The amount of signal loss depends on a flare's X-ray intensity, location of the HF path relative to the sun and design characteristics of the system. A SWF is an "immediate" effect, experienced simultaneously with observation of the causative solar flare. As a result, it is not possible to forecast a specific SWF event. Rather, forecasters can only predict the likelihood of a SWF event based on the probability of flare occurrence determined by an overall analysis of solar features and past activity. However, once a flare is observed, forecasters can quickly (within seven minutes of event onset) issue a SWF warning which contains a prediction of the frequencies to be affected and the duration of signal absorption. Normally SWFs persist only for a few minutes past the end of the causative flare, i.e., for a few tens of minutes to an hour or two.

SATCOM and Radar Interference

Solar flares can cause the amount of radio wave energy emitted by the sun to increase by a factor of tens of thousands over certain frequency bands in the VHF to SHF range (30 MHz to 30 GHz). If the sun is in the field of view of the receiver and if the burst is at the right frequency and intense enough, these radio bursts can produce direct Radio Frequency Interference (RFI) on a SATCOM link or missile detection/ space tracking radar. (Figure 3-13). Knowledge of a solar radio burst can allow a SATCOM or radar operator to isolate the RFI

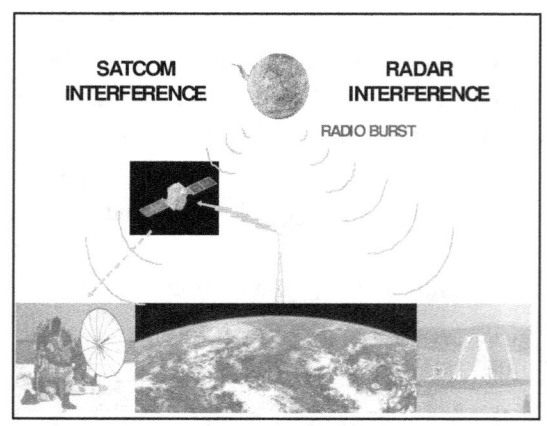

Figure 3-13: Radio Burst Effects

cause and avoid time consuming investigation of possible equipment malfunction or jamming.

Most radio signals within the 300 MHz to 300 GHz band will pass through the atmosphere, allowing communications between the ground and satellite. In the F2-region, however, radio signals can still experience some interference in the form of refraction. Refraction is the bending of the signal path just as light is refracted through a prism due to the difference in the density of the air outside the prism and the glass inside. In such a situation, the signal travels farther than just the straight-line distance between the transmitter and the receiver. This delays reception of the signal that has an impact on systems that rely on precise time, such as the Global Positioning System.

Solar radio busts are another "immediate" effect, experienced simultaneously with observation of the causative solar flare. Consequently, it is not possible to forecast the occurrence of radio bursts, let alone what frequencies they will occur on and at what intensities. Rather, forecasters can only issue rapid warnings (within 7 minutes of event onset) that identify the observed burst frequencies and intensities. Radio burst impacts are limited to the sunlit hemisphere of the Earth. They will persist only for a few minutes to tens of minutes, but usually not for the full duration of the causative flare.

There is a similar geometry-induced affect called "solar conjunction," which is when the ground antenna, satellite and the sun are in line. This accounts for why geosynchronous communication satellites will experience interference or blackouts (e.g., static or "snow" on TV signals) during brief periods on either side of the spring and autumn equinoxes. This problem does not require a solar flare to be in progress, but its effects are definitely greatest during Solar Max when the sun is a strong background radio emitter.

Sometimes a large sunspot group will produce slightly elevated radio noise levels, primarily on frequencies below 400 MHz. This noise may persist for days, occasionally interfering with communications or radar systems using an affected frequency.

Particle (Delayed) Effects

High energy particles (primarily protons, but occasionally cosmic rays) can reach the Earth within 15 minutes to a few hours after the occurrence of a strong solar flare. The event can last from hours to days. The impact of a proton event can last for a few hours to several days after the flare ends. Sample impacts include satellite disorientation, physical damage to satellites and spacecraft, false sensor readings, navigation errors and absorption of HF radio signals. Low to medium energy particle streams (composed of both protons and electrons) may arrive at the Earth about two to three days after a flare. Such particle streams can also occur at any time due to other non-flare solar activity. These particles cause geomagnetic and ionospheric storms that can last from hours to several days. Typical problems include: spacecraft electrical charging, drag on low orbiting satellites, radar interference, space tracking errors and radio wave propagation anomalies.

The sources of the charged particles (mostly protons and electrons) include solar flares, disappearing filaments, eruptive prominences and Solar Sector Boundaries (SSBs) or High Speed Streams (HSSs) in the solar wind. Except for the most energetic particle events, the charged particles tend to be guided by the interplanetary magnetic field (IMF) which lies between the sun and the Earth's magnetosphere. The intensity of a particle-induced event generally depends on the size of the solar flare, filament or prominence, its position on the sun and the structure of the intervening IMF. Alternately, the sharpness of a SSB or density/speed of a HSS will determine the intensity of a particle-induced event caused by these phenomena.

Key characteristics of particle effects are they can last for hours to days and be mostly felt in the nighttime sector (as the particles that cause them usually come from the magnetosphere's tail), although they are not limited to that time/geographic sector. Another important factor in forecasting particle events is that some of the causative phenomena (like SSBs and coronal holes, the source region for HSSs) persist for months, while the sun rotates once every 27 days. As a result, there is a tendency for these long-lasting phenomena to show a 27-day recurrence in producing geomagnetic and ionospheric disturbances.

Ionospheric Scintillation

The intense ionospheric irregularities found in the aurora zones are also one cause of ionospheric "scintillation." Scintillation of radio wave signals is the rapid, random variation in signal amplitude, phase and/or polarization caused by small scale irregularities in the electron density along a signal's path. Ionospheric radio wave scintillation is very similar to the visual twinkling of starlight or heat shimmer over a hot road caused by atmospheric turbulence. The result is signal fading and data drop-outs on satellite command uplinks, data down-links or on communications signals.

Scintillation tends to be a highly localized effect. Only if the signal path penetrates an ionospheric region where these small scale electron density irregularities are occurring will an impact be felt. Low latitude, nighttime links with geo-synchronous communications satellites are particularly vulnerable to intermittent signal loss due to scintillation. In fact, during the Persian Gulf War, allied forces relied heavily on SATCOM links, and scintillation posed an unanticipated, but very real operational problem.

Scintillation is also frequency dependent; the higher the radio frequency (all other factors held constant), the lesser the impact of scintillation. Statistically, scintillation tends to be most severe at lower latitudes (within + 20 degrees of the geomagnetic equator) due to ionospheric anomalies in that region. It is also strongest from local sunset until just after midnight, and during periods of high solar activity. At higher geomagnetic latitudes (the aurora and Polar Regions), scintillation is strong, especially at night, and its influence increases with higher levels of geomagnetic activity. Operators who know of those time periods and portions of the ionosphere where conditions are conducive to scintillation can reschedule activities or to switch to less susceptible radio frequencies.

There is no fielded network of ionospheric sensors capable of detecting real-time scintillation occurrence or distribution. Presently space environmental forecasters are heavily dependent on its known association with other environmental phenomena (such as aurora) and scintillation climatology.

GPS and Scintillation

GPS satellites, which are located at semi-synchronous altitude, are also vulnerable to ionospheric scintillation. Signal strength enhancements and fades as well as phase changes due to scintillation, can cause a GPS receiver to lose signal lock with a particular satellite. The reduction in the number of simultaneously useable GPS satellites may result in a potentially less accurate position fix. Scintillation occurrence is positively correlated with solar activity and it has been shown, under strong scintillation, that the GPS signals cannot be seen through the background noise due to the rapid changes in the ionosphere, even with the use of dual frequency receivers.

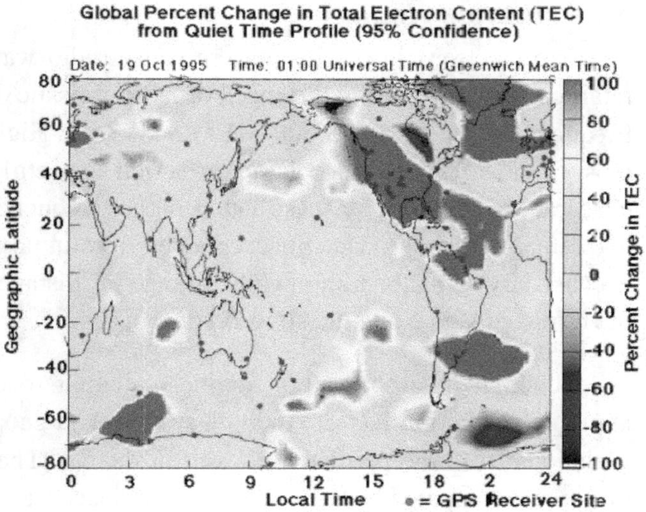

Figure 3-14: GPS and Ionospheric Disruptions

GPS and Total Electron Content (TEC)

The TEC along the path of a GPS signal can introduce a positioning error. Just as the presence of free electrons in the ionosphere caused HF radio waves to be bent (or refracted), the higher frequencies used by GPS satellites will suffer some bending (although to a much lesser extent than with HF radio waves). This signal bending increases the signal path length. In addition, passage through an ionized medium causes radio waves to be slowed (or retarded) somewhat from the speed of light. Both the longer path length and slower speed can introduce up to 300 nanoseconds (equivalent to about 100 meters) of error into a GPS location fix--unless some compensation is made for the effect. The solution is relatively simple for two-frequency GPS receivers, since signals of different frequency travel at different speeds through the same medium. Unfortunately, this approach will not work for single-frequency receivers. Thankfully, modeling inside the GPS signal can offset much of this problem. Additionally, this effect is in a limited region of the GPS receiver's field of view, so other satellites may still be able to get their signal to the receiver unaffected.

Radar Aurora Clutter and Interference

As previously discussed, a geomagnetic and ionospheric storm will cause both enhanced ionization and rapid variations (over time and space) in the degree of ionization throughout the aurora oval. Visually, this phenomenon is observed as the Aurora or Northern /Southern Lights. This enhanced, irregular ionization can also produce abnormal radar signal back-scatter on poleward looking radars, a phenomenon known as "radar aurora."

Impacts can include increased clutter and target masking, inaccurate target locations and even false target or missile launch detection. While improved software screening programs have greatly reduced the frequency of false aircraft or missile launch detection, they've not been eliminated totally.

Surveillance Radar

Free electrons in the ionosphere cause radio waves to be bent (or refracted) as well as slowed (or retarded) somewhat from the speed of light. Missile detection and spacetrack radars operate at Ultra High Frequencies (UHF, 300-3,000 MHz) and Super High Frequencies (SHF, 3,000-30,000 MHz) to escape most of the effects of ionospheric refraction so useful to HF surface-to-surface radio operators. However, even radars operating at these much higher frequencies are still susceptible to enough signal refraction and retardation to produce unacceptable errors in target bearing and range. A bearing (or direction) error is caused by signal bending, while a range (or distance) error is caused by both the longer path length for the refracted signal and the slower signal speed.

Although radar operators routinely attempt to compensate for these bearing and range errors by applying correction factors, individual solar and geophysical events will cause unanticipated, short-term variations from the predicted correction factors. These variations (which can be either higher or lower than the anticipated values) will lead to inaccurate position determinations or difficulty in acquiring targets. Real-time warnings when significant variations are occurring help radar operators minimize the impacts of their radar's degraded accuracy.

The bearing and range errors introduced by ionospheric refraction and signal retardation (as described above) also apply to space-based surveillance systems. For example, a space-based sensor attempting to lock on to a ground radio emitter may experience a geolocation error.

Atmospheric Drag

Another source for space object positioning errors is that of either more or less atmospheric drag than expected on low orbiting objects (generally at less than about 1,000 km altitude). Satellites in orbit around the Earth experience atmospheric drag. Atmospheric drag is the resistive force of the molecules of the atmosphere. The amount of drag is dependent on the density of the atmosphere, the shape and size of the object, the material it is made of, and other factors. Energy deposited in the Earth's upper atmosphere by EUV, X-ray and charged particle bombardment heats the atmosphere, causing it to expand outward. Low earth-orbiting satellites and other space objects then experience denser air and more frictional drag than expected. The expansion of the atmosphere will also result in more atmospheric drag at higher orbital altitudes thus affecting satellites that would not normally experience atmospheric drag.

This drag decreases an object's altitude and increases its orbital speed. Atmospheric drag causes a satellite to slow down, thus the satellite "falls" to a lower altitude to gain velocity. The result is the object will be some distance below and ahead of its expected position when a ground radar or optical telescope attempts to locate it. Conversely, exceptionally calm solar and/or geomagnetic conditions will cause less atmospheric drag than predicted and an object would be higher and behind where it was expected to be found.

The approximate value of atmospheric drag can be calculated and considered in updating a satellite's position and deciding on orbital maneuvers to maintain a satellite in orbit. Long term predictions are difficult because of the inability to predict what the Sun will do. The consequences of atmospheric drag include: (1) inaccurate satellite locations which can hinder rapid acquisition of SATCOM links for commanding or data transmission; (2) costly orbit maintenance maneuvers may become necessary; and

(3) de-orbit predictions may become unreliable. A classic case of the latter was SkyLab. Geomagnetic activity was so severe, for such an extended period, that the expanded atmosphere caused SkyLab to de-orbit and burn-in before a planned Space Shuttle rescue mission was ready to launch.

SPACE ENVIRONMENT EFFECTS ON SATELLITES

The environment in space has significant effect on the satellites themselves. Atmospheric drag was discussed above. The discussion below highlights the other principal effects experienced by satellites orbiting the Earth.

Radiation Hazards

Despite all engineering efforts, satellites are still quite susceptible to the charged particle environment. In fact, with newer microelectronics and their lower operating voltages, it will actually be easier to cause electrical upsets than on older, simpler vehicles. Furthermore, with the perceived lessening of the man-made nuclear threat, there has been a trend to build new satellites with less nuclear radiation hardening. This previous hardening also protected the satellites from space environmental radiation hazards.

Both low and high earth-orbiting spacecraft and satellites are subject to a number of environmental radiation hazards, such as direct physical damage and/or electrical upsets caused by charged particles. These charged particles may be: (1) trapped in the "Van Allen Radiation Belts," (2) in directed motion during a geomagnetic storm or (3) protons/cosmic rays of direct solar or galactic origin.

"Geosynchronous" orbit (35,782 km or 22,235 statute miles altitude) is commonly used for communication satellites. Unfortunately, it lies near the outer boundary of the Outer Belt of the Van Allen Radiation Belt, and suffers whenever that boundary moves inward or outward. Semi-synchronous orbit (which is used for GPS satellites) lies near the middle of the Outer Belt (in a region called the "ring current") and suffers from a variable, high density particle environment. Both orbits are particularly vulnerable to the directed motion of charged particles that occurs during geomagnetic storms. Particle densities observed by satellite sensors can increase by a factor of 10 to 1,000 over a time period as short as a few tens of minutes.

Electrical Charging

One of the most common anomalies caused by the radiation hazards discussed above is spacecraft or satellite electrical charging. Charging can be produced by: (1) an object's motion through a medium containing charged particles (called "wake charging"), which is a significant problem for large objects like the Space Shuttle or a space station, (2) direct particle bombardment, as occurs during geomagnetic storms and proton events, or (3) solar illumination, which causes electrons to escape from an object's surface (called the "photoelectric effect"). The impact of each phenomenon is strongly influenced by variations in an object's shape and the materials used in its construction

An electrical charge can be deposited either on the surface or deep within an object. Solar illumination and wake charging are surface charging phenomena. For direct particle bombardment, the higher the energy of the bombarding particles, the deeper the charge can be placed. Normally electrical charging will not (in itself) cause an electrical upset or damage. It will deposit an electrostatic charge which will stay on the vehicle (for perhaps many hours) until some triggering mechanism causes a discharge or arcing. Such mechanisms include: (1) a change in particle environment, (2) a change in solar illumination (like moving from eclipse to sunlit) or (3) on-board vehicle activity or commanding.

The two primary mechanisms responsible for charging are plasma bombardment and photoelectric effects. Plasma bombardment occurs due to varying plasma density, resulting in the surface of the satellite becoming electrostatically charged. Charging from plasma bombardment usually results in a negative charge on the surface of the satellite. The photoelectric effect results from solar radiation which liberates electrons on a satellite's surface, resulting in a positive charge on the satellite's sunlit side. A satellite will usually have a negative potential on shaded areas (due to plasma charging) and a positive potential on sunlit areas (due to the photoelectric effect). If the surface of the satellite is conductive, a current will develop to cancel these potentials. For a non-conducting surface, the charge separation will be maintained until voltage exceeds the resistive threshold of the material. This leads to a sudden electrostatic discharge.

The satellites most vulnerable to charging/discharging are those located at geosynchronous altitude; discharges as high as 20,000 volts (V) have been experienced. Satellites in geosynchronous orbits typically move both in and out of the upper regions of the Van Allen Radiation Belts and the Earth's magnetotail. A single proton or cosmic ray can (by itself) deposit enough charge to cause an electrical upset. Sudden electrostatic discharge (high current or arc) can cause hardware damage, such as: (1) blown fuses or exploded transistors, capacitors and other electronic components, (2) vaporized metal parts and structural damage, and (3) breakdown of thermal coatings. They can cause electrical or electronic problems, such as: (1) false commands, (2) spurious circuit switching, (3) memory changes, (4) solar cell degradation, and (4) false sensor readings. Warnings of environmental conditions conducive to spacecraft charging allow operators to reschedule vehicle commanding, reduce on-board activity, delay satellite launches and deployments or re-orient a spacecraft to protect it from particle bombardment.

Outgassing

Although the environment in space is not benign, the density of particles above 100 miles altitude is extremely low. There is almost no atmospheric pressure, similar to a complete vacuum. As a result, satellites and the materials they are made of experience phenomena which are not encountered on Earth. In a vacuum, some materials experience outgassing. Outgassing is a phenomenon where molecules of material evaporate into space. Although many materials experience outgassing, composite materials and those made with volatile solvents are particularly susceptible. These include electronic microchips, plastics, glues and adhesives. Outgassing can result in changes to the physical properties of a material. In addition, the evaporating molecules can form a thin film over other components of the satellite, thereby affecting their performance. Outgassing can be minimized through careful selection of component materials but eventually some components will exhibit different characteristics and properties.

OTHER SPACE PROBLEMS

Space Debris

Space debris is defined as any non-operational man-made object of any size in space. The size of space debris varies from complete inoperative satellites and expended rocket bodies to small chips of paint. Of the more than 12,000 man-made items in space currently tracked and catalogued, only about 22% are operational space systems. The rest is space debris. Space debris smaller than approximately 2 cm (0.78 inches) cannot be detected and tracked reliably, therefore it is reasonable to assume that there is significantly more space debris than we know about. It has been estimated that as many as 100 satellites have broken up while in orbit, sometimes due to explosions of propulsion systems, and at other times due to impact with other space debris. The result is an estimated 40,000 to 80,000 pieces of debris in orbit around the Earth. There is even a wrench that became space debris when it drifted away from an astronaut during a space walk. Most debris is small but it can be traveling at relatively high speeds.

Figure 3-15: Space Debris Cartoon

Debris in low Earth orbit tends to have much higher velocities relative to other objects in orbit than those at geostationary orbit altitude (22,300 miles). In a collision between two objects, the relative velocity and the masses of the objects determine the force of impact. A dense object like the wrench could do catastrophic damage if it were to hit a satellite at even a low relative velocity. At high impact velocities of 30,000 mph, even small objects are capable of inflicting significant damage. A small paint chip damaged a window on the space shuttle when it impacted at about 8,000 miles per hour. Shielding, energy absorbing panels and other design considerations can make a satellite more resistant to damage from impacts with small space debris. At altitudes below 200 miles, atmospheric drag tends to cause much debris to reenter the Earth's atmosphere where it is usually vaporized. Space debris located at the altitude of geostationary orbits tends to have lower relative velocities (200 - 1,000 mph) and the density is much less. Atmospheric drag at this altitude is almost zero; therefore the debris is present for a much longer time.

Meteoroids

It is estimated that approximately 20,000 tons of natural material is added to the Earth each year from impacts of meteoroids and asteroid fragments with the Earth's atmosphere. Most of these particles are the size of dust particles; however, some are much larger. When meteoroids enter the Earth's atmosphere they usually burn up due to the friction with the air molecules. Larger meteoroids generate enough light to be seen as meteors streaking across the night sky. Occasionally, larger objects don't completely vaporize. When a piece strikes the surface of the Earth it is called a meteorite. These particles represent a constant natural danger to satellites in orbit around the Earth. The Long Duration Exposure Facility (LDEF) was a US research satellite that remained in orbit for six years before it was recovered by the Shuttle and returned to the Earth. Examination of the exposed surfaces indicates thousands of impacts by micro-meteoroids. Microscopic examination has revealed extensive damage to metal surfaces. Most meteoroids are too small and traveling too fast to be detectable in time for satellite controllers to direct a satellite to change it's orbit to avoid collision. Shielding and other design considerations are the most effective means to protect satellites from catastrophic damage.

An example of concern would be from the comet Temple-Tuttle in November 1998. Meteoroids normally travel at about 12 miles per second; but because of the relationship of the Earth's orbit to Temple-Tuttle's orbit, Leonids meteoroids rocketed past at the speed of a 22-caliber bullet. If an impact occurs; if the resulting impact causes a discharge; if the discharge gets inside the satellite vs. escaping into the space environment; if the path of that discharge hits an electrical component, then the result could be a fried satellite. To minimize the storm's damage, space personnel can determined a comprehensive series of mitigation strategies to protect space assets and allow the Air Force to continue its vital missions. These strategies included normal precautions such as powering down unnecessary onboard electronics and reducing a satellite's cross-section.

SPACE ENVIRONMENT REFERENCES

Websites:

http://umtof.umd.edu/pm/ - This site provides a real time printout of the last 48 hours of solar wind activity.

http://www.ngdc.noaa.gov/stp/GLOSSARY/glossary.html -Index of solar and terrestrial weather terms.

http://hesperia.gsfc.nasa.gov/hessi/brochure.htm - HESSI, High Energy Solar Spectroscopic Imager, at NASA Goddard Space Flight Center

http://lheawww.gsfc.nasa.gov/docs and http://universe.gsfc.nasa.gov/ - NASA Educational sites for Space Weather basic information among other things

http://hesperia.gsfc.nasa.gov/hessi/flares.htm - A solar flare is an enormous explosion in the solar atmosphere, involving sudden bursts of particle acceleration, plasma heating, and bulk mass motion. It is believed to result from the sudden release of energy stored in the magnetic fields that thread the solar corona in active regions around sunspots

http://solar.uleth.ca/ - This site provides information on the state of the sun and its affects on earth.

http://solarscience.msfc.nasa.gov/SunspotCycle.shtml - Up-to-date monitoring of sunspots and the solar cycle, with tutorials, material for educators, and interactive activities.

http://hesperia.gsfc.nasa.gov/sftheory/ - These pages are about solar flares, the biggest explosions in the solar system. Their purpose is to provide some general information about solar flares.

http://www.spacescience.org/ExploringSpace/SpaceWeather/1.html – Good tutorial of how the sun actually impacts us here on Earth.

http://spaceweather.com/ - Site to see current conditions.

http://solarham.com/ - Site tracking current conditions as a compendium of other sites.

Chapter 4
Satellite Orbits

OVERVIEW

Satellites move predictably according to the laws of physics. Satellites do not escape Earth's gravity; in fact, it is Earth's gravity that holds them in orbit. Without gravity there would not be any satellites because nothing would stay in orbit. Many different types of orbits can be achieved by changing a satellite's orbital parameters. For example, it is possible to position a satellite over the equator at an altitude where the time for it to revolve around the Earth is exactly one day. The result is the satellite will appear to be stationary to an observer on the surface of the rotating Earth below. It is also possible to create an orbit so that a satellite will pass within view of every point on the Earth at the same time every day and night. Some orbits are not possible. For example, it is not possible to position a satellite in low Earth orbit so that it will loiter or hover over a particular spot. The orbit chosen for a satellite is determined by the mission it is designed to perform.

The intent of this section is to present information in non-mathematical terms that allow the reader to develop an understanding of the motion of orbiting satellites. This provides a basis for understanding the capabilities and limitations of various orbits, particularly with respect to the service that the satellite provides to the user.

BASICS PHYSICS OF ORBITS

Kepler's Laws of Planetary Motion

In the early 1600's, Johannes Kepler formulated three Laws of Planetary Motion. Although his laws were intended to explain the motion of the planets around the Sun, they also apply to any satellite orbiting around another object.

Kepler's First Law: Law of Ellipses

"The orbit of each planet is an ellipse with the sun at one focus."

As it applies to satellites in orbit around the Earth, Kepler's first law can be restated as,

"The orbit of each satellite is an ellipse with the center of the Earth at one focus." (Figure 4-1)

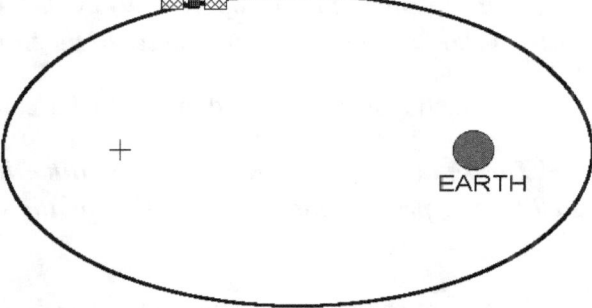

Figure 4-1: Kepler's First Law

Neglecting such influences such as atmospheric drag, mass asymmetry, and third body effects, the law applies accurately to all orbiting bodies. With Kepler's second law, he was on the trail of Newton's Law of Universal Gravitation. He was also hinting at calculus, which was not yet invented.

Kepler's Second Law: Law of Areas

"Every planet revolves so that the line joining it to the sun sweeps over equal areas in equal times anywhere in the orbit."

For earth orbiting satellites this can be restated as,

"A satellite orbits so that the line joining it with the center of Earth sweeps over equal areas in equal times anywhere in the orbit."

In a circular orbit a satellite travels at the same speed at all points. In an elliptical orbit the speed of the satellite varies. As the satellite approaches closer to the Earth its speed increases. The maximum speed is attained at the point of closest approach called the perigee. As the satellite moves away from the Earth it slows down. The slowest velocity occurs at the point farthest from the Earth, the apogee.

Kepler's Third Law: Law of Harmonics

Kepler discovered his third law ten years after he published the first two in <u>Astronomia Nova</u> (*New Astronomy*). He had been searching for a relationship between a planet's *period* and its *distance* from the Sun since his youth. Kepler was looking at harmonic relationships in an attempt to explain the relative planetary spacing. After many false steps and dogged persistence, he fell upon his famous relationship: Kepler's Third Law: Law of Harmonics

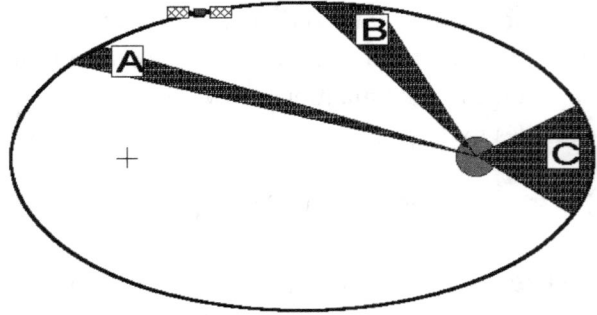

AREA A = AREA B = AREA C

Figure 4- 2: Kepler's Second Law

"The square of the sidereal periods (the time it takes to complete one orbit of the Sun) of any two planets are to each other (proportional) as the cube of their mean distances from the center of the Sun."

For a satellite in orbit around the Earth this can be restated as,

"The squares of the orbital periods (time it takes a satellite to complete one orbit) of any two satellites are proportional to each other as the cubes of their mean distances from the center of the Earth."

The square of the period divided by the cube of the mean distance from the center of the Earth is a constant for all objects. The constant number is not dependent upon the mass of the satellite.

Kepler's third law of planetary motion has many implications. First, satellites of different masses with exactly equal orbits have equal speeds. The mass of a satellite does not determine its period or orbital speed. Second, satellites with orbits of equal semi-major axis length have equal periods, whether or not their eccentricities are the same. Finally, comparing two satellites with orbits of equal eccentricity, the orbit with the longer semi-major axis has a larger circumference, and the satellite in the larger circumference orbit has a lower speed. Thus, satellites farther away from the Earth have more distance to travel and move more slowly than satellites closer to the Earth.

Newton's Laws

Sir Isaac Newton was a British scientist who, in the late 1600's, formulated his laws of motion and a law of universal gravitation. Kepler's laws provided mathematical formulas for the orbits of satellites but did not explain what forces cause satellites to move in space the way that they do. Many great thinkers were on the edge of discovery, but it was Newton that took the pieces and formulated a grand view that was consistent and capable of describing and unifying the mundane motion of a "falling apple" and the motion of the planets:

Newton's First Law: Law of Inertia

"Every body continues in a state of rest or of uniform motion in a straight line unless it is compelled to change that state by a force impressed on it."

An object at rest will remain at rest and an object in motion will remain in motion in a straight line, unless it is acted upon by an outside force. If an object deviates from rest or motion in a straight line with constant speed, then some force is being applied. Newton's First Law describes undisturbed motion; inertia, accordingly, is the resistance of mass to changes in its motion. His Second Law describes how motion changes. It describes the relationship between the impressed forces, the masses of objects, and the resulting motion:

Newton's Second Law: Law of Momentum

"When a force is applied to a body, the time rate of change of its momentum is proportional to, and in the direction of, the applied force."

Momentum is a measure of an object's motion. It is the property of a moving body that determines what force is required to bring the body to rest in a given amount of time. Newton continued his discoveries and with his third law, completed his grand view of motion:

Newton's Third Law: Law of Action-Reaction

"For every action there is a reaction, equal in magnitude, but opposite in direction, to the action."

The effect of this law is most easily demonstrated where friction is not present or is extremely low. This is the law that explains how rockets work. The rocket exhaust gases are light but are ejected at high velocity. This results in a force that accelerates the larger mass of the rocket. The maximum velocity of the rocket is the maximum velocity of the rocket gases; however, the propellant is usually expended before that velocity is attained.

Newton's Law of Universal Gravitation

Newton theorized gravity, which he believed to be responsible for the "falling apples" and the planetary motion. He formulated the Law of Universal Gravitation. It states,

"Between any two objects in the universe there exists a force of attraction that is in proportion to the product of the masses of the objects and is in inverse proportion to the square of the distance between them."

Every particle in the universe attracts every other particle with a force that is proportional to the product of the masses and inversely proportional to the square of the distance between the particles.

$$F = G \, m_1 m_2 / d^2$$

Where F is the force due to gravity, G is the proportionality constant, m_1 and m_2 the masses of the central and orbiting bodies, and d the distance between the two bodies.

The gravitational attraction between two objects is greater for more massive objects and decreases as the square of the distance between the objects increases. "G" is the Universal Gravitational Constant which is the same everywhere in the universe. Newton based his derivation on large objects rotating mutually about each other, such as the Earth and the Moon. It applies equally to man-made satellites orbiting the Earth except that the mass of the man-made satellite is so small compared to the mass of the Earth that the offset of the center of rotation of the two objects from the center of the Earth's mass is so small it is ignored.

It is the force of gravity that allows satellites to orbit. Gravity is the force that continually acts on the satellite to continually change its direction in order to circle Earth. Without gravity, the satellite would fly off into space in a straight line, maintaining constant velocity.

The Law of Universal Gravitation states that all objects with mass in the universe exert a gravitational attraction on all other objects. The number of objects in the universe is almost infinite but the gravitational forces from objects farther away than the immediate solar system are very insignificant. The force of the Earth's gravity at sea level results in an acceleration of about 32 ft/sec^2. This is called a "1 g force". For a satellite in orbit 200 miles above the Earth the effects of gravity from the most significant sources are shown below:

Gravitational Effect of an Earth Satellite with 200 mile Altitude		
Source of Gravitation	Strength of Gravitational Force	
Earth	0.89 g	28.600 ft/sec^2
Earth's Oblateness	0.001 g	0.030 ft/sec^2
Sun	0.0006 g	0.019 ft/sec^2
Moon	0.00000033 g	0.001 ft/sec^2

Table 4-1: Gravitational Effects

Satellites do not escape Earth's gravity but are held in orbit by it. Without the constant force of Earth's gravitational attraction on the satellite it would go off into deep space in a straight line. As the distance from the Earth increases the Earth's gravitational force decreases, thus the gravitational effects of other bodies in the solar systems become relatively more significant.

A misconception is that centrifugal force somehow affects the orbit of a satellite. It does not. Centrifugal force only exists when there is a physical link between two objects and one is moving around the other. A ball tied to the end of a string and swung around in a circle is an example. If the string breaks, the ball moves in a straight line except that the Earth's gravity pulls it toward the center of the

Earth. The ball follows an arching path until it hits the ground. Obviously, there is no string, rope or equivalent attached to each satellite. The principle force that holds satellites in orbit is gravity.

Newton's Derivation of Kepler's Laws

Kepler's laws of planetary motion are empirical (found by comparing vast amounts of data in order to find the algebraic relationship between them); they describe the way the planets are observed to behave. Newton proposed his laws as a basis for all mechanics. Thus Newton should have been able to derive Kepler's laws from his own, and he did:

Newton's Derivation of Kepler's First Law: If two bodies interact gravitationally, each will describe an orbit that can be represented by a conic section about the common center of mass of the pair. In particular, if the bodies are permanently associated, their orbits will be ellipses. If they are not permanently associated, their orbits will be hyperbolas.

Newton's Derivation of Kepler's Second Law: If two bodies revolve about each other under the influence of a central force (whether they are in a closed orbit or not), a line joining them sweeps out equal areas in the orbit plane in equal intervals of time

Newton's Derivation of Kepler's Third Law: If two bodies revolve mutually about each other, the sum of their masses times the square of their period of mutual revolution is in proportion to the cube of their semi-major axis of the relative orbit of one about the other.

ORBITAL MOTION

Newton's laws of motion apply to all bodies, whether they are scurrying across the face of the Earth or out in the vastness of space. By applying Newton's laws one can predict macroscopic events with great accuracy. According to Newton's first law, bodies remain in uniform motion unless acted upon by an external force; that uniform motion is in a straight line. This motion is known as inertial motion, referring to the property of *inertia*, which the first law describes.

Velocity is a relative measure of motion. While standing on the surface of the Earth, it seems as though the buildings, rocks, mountains and trees are all motionless; however, all of these objects are moving with respect to many other objects (Sun, Moon, stars, planets, etc.). Objects at the equator are traveling around the Earth's axis at approximately 1,000 mph; the Earth and Moon system is traveling around the Sun at 66,000 mph; the solar system is traveling around the galactic center at approximately 250,000 mph, and so on and so forth.

To put a man-made object into orbit around the Earth requires that a number of conditions be met. The most critical are velocity and altitude. According to Newton's second law, for a body to change its motion there must be a force imposed upon it. Everyone has experience with changing objects' motion or compensating for forces that change their motion. An example is playing catch; when throwing or catching a ball, its motion is altered; thus, gravity is compensated for by throwing the ball upward by some angle allowing gravity to pull it down, resulting in an arc. If the ball is initially motionless, it will fall straight down. However, if the ball has some horizontal motion, it will continue in that motion while accelerating toward the ground. How far the ball travels along the ground in that one second depends on its horizontal velocity. At sea level an object falls 16 feet in the first second of travel due to gravity. The curvature of the Earth is about 16 feet every five miles. The faster the horizontal velocity the ball goes then the greater the horizontal distance that it must travel. Eventually one would come to the point where the Earth's surface drops away as fast as the ball drops toward it, assuming no air resistance to continually slow the ball down

To put a satellite into orbit around the Earth the satellite must be accelerated to a velocity parallel to the surface so that the curvature of the Earth compensates for the distance the satellite falls toward the center of the Earth. Near the surface of the Earth an object would have to travel about five miles per second (about 18,000 miles per hours). Of course, friction from the atmosphere near the Earth's surface is substantial. At 18,000 miles per hour air friction would quickly heat the object to a high temperature and it would burn up before traveling very far. To avoid air friction, the object must be boosted to an altitude high enough to avoid most of the atmosphere. These factors determine the basic altitude and velocity necessary to put an object into orbit around the Earth. The minimum speed for an object to be placed into orbit around the Earth is about 17,500 miles per hour and the minimum altitude of a circular orbit is about 90 miles. At this altitude the atmospheric density is still enough to cause the satellite to slow down so that it would fall out of orbit after only a couple of revolutions. It would either burn up during the high-speed reentry through the atmosphere or it would impact on the surface.

Figure 4-3 shows how differing velocity affects a satellite's trajectory or orbital path. The figure depicts a satellite at an altitude of one Earth radius (6378 km above the Earth's surface). At this distance, a satellite would have to travel at 5.59 km/sec to maintain a circular orbit and this speed is known as its *circular speed* for this altitude. As the satellite's speed increases, it falls farther and farther away from the Earth and its trajectory becomes an elongating ellipse until the speed reaches 7.91 km/sec. At this

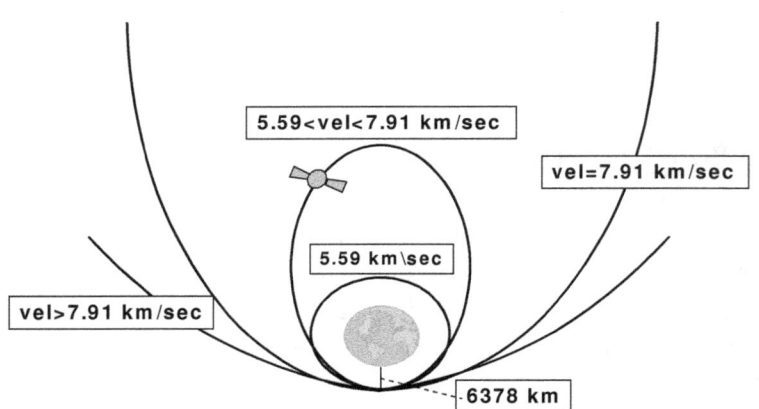

Figure 4-3: Velocity vs. Trajectory

speed and altitude the satellite has enough energy to leave the Earth's gravity and never return; its trajectory has now become a parabola, and this speed is known as its *escape speed* for this altitude. As the satellite's speed continues to increase beyond escape speed its trajectory becomes a flattened hyperbola. From a low Earth orbit of about 100 miles, the escape velocity becomes 11.2 km/sec. In the above description, the two specific speeds (5.59 km/sec and 7.91 km/sec) correspond to the circular and escape speeds for the specific altitude of one Earth radius.

Newton's three laws and his Law of Universal Gravitation describe the satellite's motion. The Law of Universal Gravitation describes how the force between objects decreases with the square of the distance between the objects. As the altitude increases, the force of gravity rapidly decreases, and therefore the satellite can travel slower and still maintain a circular orbit. For the object to escape the Earth, it has to have enough kinetic energy (kinetic energy is proportional to the square of velocity) to overcome the gravitational potential energy of its position. Since gravitational potential energy is proportional to the distance between the objects, the farther the object is from the Earth, the less potential energy the satellite must overcome, which also means the less kinetic energy is needed.

Location Reference Systems

Reference systems are used everyday. To give or follow directions both the giver and acceptor have to agree on a common reference system or the directions are worthless; left, right, north, south, the origin and so on, must be agreed upon. Once a common reference has been determined, spatial information can be traded. The same must be done when considering orbits and satellite positions. Before positions can be defined, a common reference must be agreed upon and understood. The reference system utilized depends on the situation, or the nature of the knowledge to be retrieved. Of the many reference systems available, the Geographic Coordinate Reference Systems (often referred to as the Earth-Fixed Greenwich Reference Systems) and the Geocentric (or Earth Centered) Inertial Coordinate System stand out because of their use in locating points on the surface of the Earth or describing the motion of Earth orbiting objects.

Although each location reference system has its own unique terminology, all have a fundamental plane, a point of origin or a reference point located on the fundamental plane, and a principle direction on the fundamental plane. Once the fundamental plane, origin, and principle direction have been established, it is possible to make distance and angular measurements from them. These measurements are sometimes called displacements.

Geographic Coordinate Reference Systems

These systems are used to describe locations on most maps of the Earth's surface (**Figure 4-4**). The fundamental plane is the equatorial plane of the Earth. The point of origin is the center of the Earth. The principal direction is a line from the center of the Earth through the point where the prime meridian intersects the fundamental plane. The prime meridian is 0° longitude, an imaginary line on the surface of the Earth from the North Pole to the South Pole that passes through Greenwich, England. Altitude is usually stated as distance above sea level. Sea level is approximately one Earth radius from the

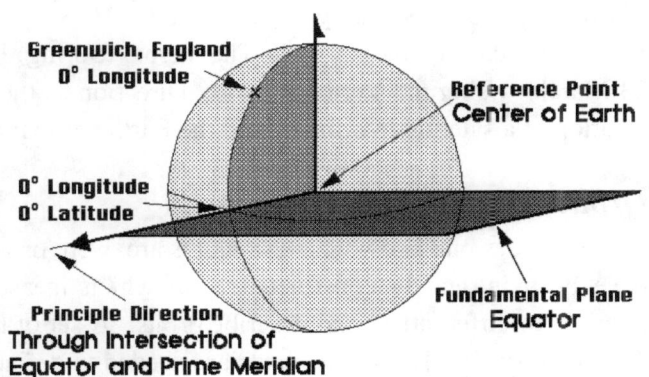

Figure 4- 4: Geographic Coordinate Reference System

center of the Earth (point of origin). The Military Grid Reference System (MGRS), the Universal Transverse Mercator (UTM) coordinate system and latitude/longitude are examples of Geographic Coordinate Reference Systems. The Earth is constantly spinning about its axis; therefore the principle direction also rotates. This makes these location reference systems unsuitable for describing the location and motion of objects orbiting the Earth.

Geocentric Inertial Coordinate System

The Geocentric (Earth-Centered) Inertial Coordinate System (**Figure 4-5**) is used to determine the orientation of an object orbiting the Earth in space. The origin is the Earth's center, the fundamental plane is the equatorial plane and the principal direction is a line to the Vernal Equinox Direction (formerly the First Point of Aries) which is the direction from the center of the Earth to the center of the Sun when the Earth is at the vernal equinox (**Figure 4-6**). This can also be drawn as a line from the vernal (spring) equinox to the autumnal (fall) equinox. The Earth rotates daily about its axis at an angle (23° 7') to the orbital (ecliptic) plane of the Earth as it rotates yearly about the Sun, thus the Earth's equatorial plane is at an angle to the ecliptic plane. Twice a year the Earth's orbit crosses the point where the ecliptic plane and the Earth's equatorial plane intersect. These points are called equinoxes because the length of daylight and night are the same. The Autumnal (Fall) Equinox normally occurs on 21

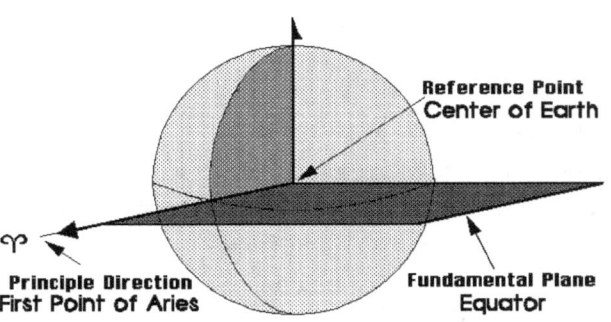

Figure 4-5: Geocentric Inertial Coordinate System

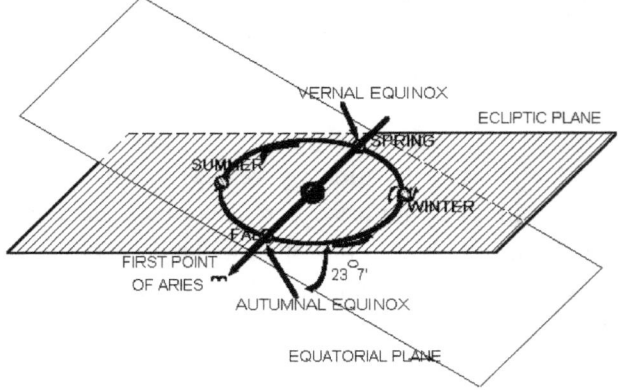

Figure 4-6: Determination of First Point of Aries

September, the first day of Fall. The Vernal (Spring) Equinox normally occurs on the first day of spring, 23 March. Using the Vernal Equinox Direction as the principle direction provides a way to describe the location of a satellite in orbit around the Earth without regard to the Earth's rotation about its axis.

Orbital Element Set

How does one know where satellites are, were or will be? Intermeshing coordinate reference systems have been intricately constructed from which measurements are defined to result in the parameters required to differentiate and describe orbits. A set of these parameters is a satellite's *orbital element set*. As stated previously, two elements are needed to define an orbit: a satellite's position and velocity. Given these two parameters, a satellite's past and future position and velocity may be predicted.

A set of orbital parameters or elements is used to describe the size and shape of a satellite's orbit, the orientation of the orbital plane, the orientation of the orbit within the orbital plane, and to locate the satellite within the orbit at any specified time. Element sets or "elsets" are used to calculate where a satellite is at a specified time and to predict where it will be in the future. Predictions are only approximations because many of the minor forces which influence a satellite's orbit are not included in the calculations. If the values in the element set are current, the predictions can be quite accurate. The values in the orbital element set must be updated more frequently for satellites in low Earth orbit than for those in geostationary orbit.

In three-dimensional spaces, it takes three parameters each to describe position and velocity. Therefore, any element set defining a satellite's orbital motion requires at least six parameters to fully describe that motion. There are different types of element sets, depending on the use. We are interested in using the Keplerian, or classical, element set. The orbital elements tell us four things we want to know about orbits, namely:

- Orbit size (semi-major axis)

- Orbit shape (eccentricity)

- Orientation (inclination)

  - Orbit plane in space

  - Orbit within plane

- Location of the satellite

Semi-Major Axis (a)

The first parameter describes an orbit's size. The maximum distance from the center of an ellipse to a point on the ellipse is called the semi-major axis, *a*. The minimum distance from the center to a point on the ellipse is called the semi-minor axis, *b*. The semi-major axis is a significant measurement since it also equals the average radius and thus, is a measure of the mechanical energy of the orbiting object. (**Figure 4-7**)

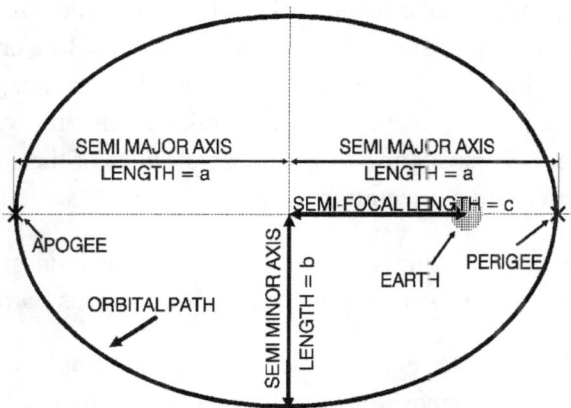

Figure 4-7: Components of and Ellipse

Eccentricity (e)

Eccentricity, *e,* measures the shape of an orbit. The shape determines the positional relationship to the central body, because the central body must occupy one of the foci of the ellipse (or other conic section). Eccentricity describes the amount an ellipse deviates from a circular shape. Eccentricity is the ratio of the distance from a focus to the center of the ellipse divided by the length of the semi-major axis.

$$e=c/a$$

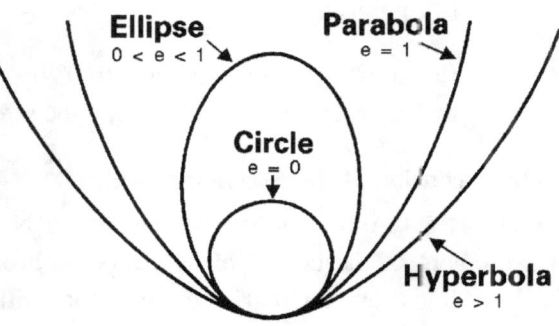

Figure 4-8: Values of Eccentricity (e)

The center of a circle is the same point as the focus; therefore the eccentricity of a circle equals 0. All elliptical orbits have an eccentricity equal to or greater than 0 but less than 1. An eccentricity equal to 1 describes a parabolic path. An eccentricity greater than 1 describes a hyperbolic path which does not represent an orbit. (**Figure 4-8.**)

Inclination (*i*)

The first angle used to orient the orbital plane is *inclination,* (**i**): a measurement of the orbital plane's tilt. Because it defines the tilt of the orbital plane, it defines the maximum latitude, both North and South, which will be directly beneath the satellite. This is an angular measurement from the equatorial plane to the orbital plane $(0° \le i \le 180°)$, measured counter-clockwise at the ascending node while looking toward Earth. The ascending node is the point in the orbit where the satellite crosses from the Southern Hemisphere to the Northern Hemisphere. The value of the inclination of an orbit is also the value of the

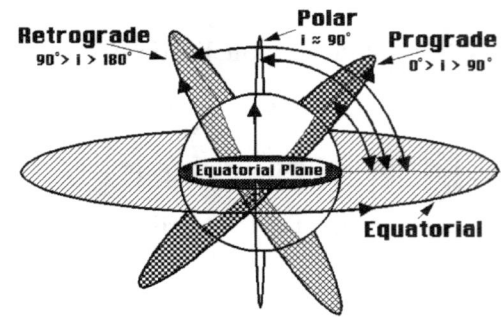

Figure 4-9: Inclination of Orbits

northernmost and southernmost latitudes of the satellite ground track. The inclination of an orbit helps to describe its orientation in space. Orbits are sometimes described by their inclination. (**Figure 4-9.**)

- Prograde (or posigrade): Inclination greater than or equal to 0° but less than 90°; the satellite orbits Earth in the same direction as Earth's rotation.

- Retrograde: Inclination greater than 90° but less than or equal to 180°; the satellite orbits Earth opposite to Earth's rotation. Sun synchronous orbits, common for weather and remote sensing satellites, are retrograde orbits.

- Polar: Inclination equal to 90°; the satellite orbits from the north pole to the south pole and back to the north pole. Orbits with an inclination very close to 90° (88° - 92°) are often referred to as being polar.

- Equatorial: Inclination equal to 0° or 180°; the satellite's orbital plane and the equatorial plane of the Earth are the same therefore, the satellite is in orbit directly above the equator of Earth.

Right Ascension of the Ascending Node (Ω)

The Right Ascension of the Ascending Node is a measurement of the orbital plane's rotation around the Earth. It is an angular measurement within the equatorial plane from the Vernal Equinox Direction eastward to the ascending node $(0° \le \Omega \le 360°)$ (**Figure 4-10**). The First Point of Aries is not a fixed point in space although it has been described as such: it actually

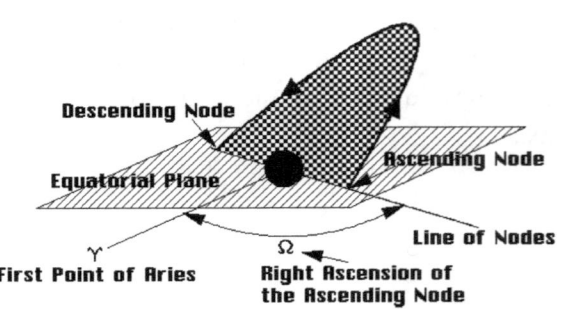

Figure4-10: Right Ascension of the Ascending Node

moves along the Ecliptic at a rate of roughly one degree every seventy years. When the Equinox was first observed, thousands of years ago, the First Point actually lay in the constellation of Aries. Due to the effect of precession, the First Point of Aries crossed into the neighboring constellation of Pisces in about 70 BCE. It has taken about 2,000 years to cross Pisces, and it will cross into the next zodiacal constellation, Aquarius, in about the year 2600. Following its journey along the Ecliptic, it will return to Aries once again in about 23,000 years. That is why orientation is now from the Vernal Equinox Direction.

The ascending node is the point where the satellite's orbit intersects the Earth's equatorial plane when the satellite is crossing from south to north. The line of nodes is an imaginary line between the ascending node and the descending node. Since an orbital plane always intersects the center of the Earth, the line of nodes will always pass through the center of the Earth. The right ascension of the ascending node helps to describe the orientation of a satellite's orbit in space.

Argument of Perigee (ω)

Inclination and Right Ascension fix the orbital plane in inertial space. The orbit must now be fixed within the orbital plane. For elliptical orbits, the perigee is described with respect to inertial space. The *Argument of Perigee (ω)* orients the orbit within the orbital plane. It is an angular measurement within the orbital plane from the ascending node to perigee in the direction of satellite motion $(0° \leq \omega \leq 360°)$

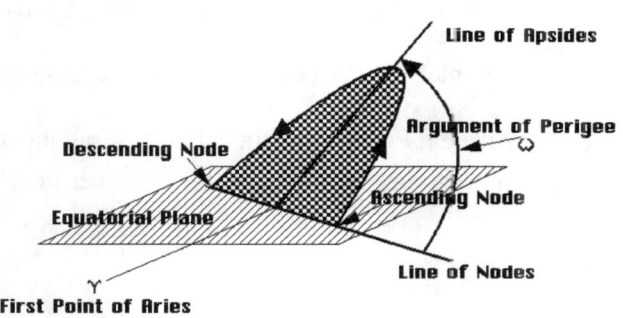

Figure 4-11: Argument of Perigee

(**Figure 4-11**). The apogee is directly opposite the perigee, a difference of 180°. The line of apsides is an imaginary line between the apogee and the perigee of an orbit. It will also pass through the center of the Earth. The argument of perigee helps to describe the orientation of a satellite's orbit in its orbital plane.

True Anomaly (ν)

At this point all the orbital parameters needed to visualize the orbit in inertial space have been specified. The final step is to locate the satellite within its orbit. True anomaly is the angular measurement (0° to 360°) from the center of the Earth in the direction of a satellite's motion from the point of perigee along the orbital path to the location of the satellite at a specific or Epoch time. Epoch time is an arbitrary time used as a reference point at which other orbital parameters are measured. The Air Force Space Command states the epoch time when the satellite is at the ascending node. In traditional Keplerian element sets, epoch time is usually stated when the satellite is at its perigee. An orbit number or revolution number associated with the epoch time is usually included in the orbital element set.

Mean Anomaly

The mean anomaly is the angle measured from the center of an ellipse from the perigee to the satellite's position through which it would move in a specified period of time if it moved at its mean angular rate of motion. The mean angular rate of motion is the period of the orbit divided by 360° (one complete revolution). The mean anomaly is the same as the true anomaly for a satellite in a perfectly circular orbit.

Period

The element set describes the average orbit for a satellite. Due to perturbations, the satellite's actual orbit will deviate somewhat from a perfect ellipse. The period defines the ground track's *westward regression*. The period of a satellite is not an orbital element, but it is extremely useful. There are four different ways to describe the period of a satellite:

- Nodal Period: The time it takes a satellite to travel once from ascending node to ascending node. In most space discussions, "period" is used when referring to the nodal period.

- Keplerian Period: The period defined by Kepler's third law. This period does not take perturbations into account and is only an average.

- Anomalistic Period: The time it takes a satellite to travel once from perigee to perigee.

- Sidereal Period: The time it takes a satellite to cross the equatorial plane at 0° right ascension from the last time it crossed at 0° right ascension.

Element	Name	Description	Definition	Remarks
a	semi-major axis	Orbit size	half of the long axis of the ellipse	orbital period and energy depend on orbit size
e	eccentricity	Orbit shape	ratio of half the foci separation (c) to the semi-major axis	closed orbits: $0 \le e < 1$ open orbits:$1 \le e$
i	inclination	Orbital plane's tilt	angle between the orbital plane and the equatorial plane, measured counter clockwise at the ascending node	equatorial: $i = 0°$, 180° prograde:$0° \le i < 90°$ polar: $i = 90°$ retrograde: $90° < I \le 180°$
Ω	right ascension of the ascending node	Orbital plane's rotation about the Earth	angle, measured eastward, from the vernal equinox to the ascending node	$0° \le \Omega < 360°$ undefined when $i = 0°$, 180° (equatorial orbit)
ω	argument of perigee	Orbit's orientation in the orbital plane	angle, measured in the direction of satellite motion, from the ascending node to perigee	$0° \le \omega < 360°$ undefined when $i = 0°$, 180°, or $e = 0$ (circular orbit)
ν	true anomaly	Satellite's location in its orbit	angle, measured in the direction of satellite motion, perigee to satellite's location	$0° \le \nu < 360°$ undefined when $e = 0$ (circular orbit)

Table 4-2: Classical Orbital Elements

TYPES OF ORBITS

Low Earth Orbits

There is no formal definition of what constitutes a Low Earth Orbit (LEO) but it is generally considered to have an apogee (maximum altitude) of no more than approximately 530 miles (830 km). Inclination of the orbital plane can be any value. Most low Earth orbits are nearly circular, therefore the eccentricity is very close to zero. At low altitudes, atmospheric drag significantly limits the lifetime of satellites unless they are periodically boosted into a higher orbit. Orbital lifetime, without any propulsion, is about one year at an altitude of 200 miles (320 km). Orbital lifetime at 500 miles (805 km) altitude is more than 10 years. A considerable amount of space debris has collected in the higher altitudes, thus increasing the chances of having a collision with a piece of space debris or a meteoroid large enough to do significant damage to an orbiting satellite.

Low Earth orbits are commonly used for observation, environmental monitoring, small communications satellites, and science instrument payloads. Manned orbiting satellites, such as the US Space Shuttle and the International Space Station (ISS), generally remain below an altitude of 300 miles (483 km) so that heavy shielding to protect the crew from radiation from the Van Allen radiation belts is not needed.

Satellites in LEO have the advantage that they pass relatively close to areas on the Earth. This makes them significant to those who image the earth from space. Also, this orbit is good if you wish to keep antenna size and power down. The closer the satellite is to the earth the smaller the antenna can be to receive the signal and the less power required to transmit to and from the satellite. Surveillance and some weather satellites use this type of orbit.

There are disadvantages to this type of orbit which must be considered. The satellite is in view of the ground user for only a short period of time as it passes quickly overhead and the footprint of the satellite can be quite small. A satellite in LEO cannot provide continuous coverage to a specific geographic point or area. The orbit *period* at these altitudes varies between ninety minutes and two hours. The radius of the *footprint* of a communications satellite in LEO varies from 1864 to 2485 miles (3000 to 4000 km). The maximum time during which a satellite in LEO orbit is above the local horizon is about 20 minutes for an observer on the earth. In many orbits, the satellite may orbit the Earth many times before it passes within view again. A global communications system using this type of orbit requires a large number of satellites, in a number of different, inclined, orbits. When a satellite serving a particular user moves below the local horizon, it needs to be able to hand over the service to a succeeding one in the same or adjacent orbit. Lastly, maintaining constant communications with a low Earth orbiting satellite requires a large number of ground stations located around the world or a few relay satellites in high, geostationary orbit. The shuttle and Landsat satellites communicate through relay satellites.

"Big LEO" systems are normally those providing mainly mobile telephone services. Many of the new proposed 'global mobile phone' services will be provided by this type of satellite. They are located between 435 – 932 miles (700-1,500 km) from the Earth. Examples of Big LEO" systems include Globalstar (designed for 48+8 satellites in 8 orbital planes at 870 miles (1400 km)), and Iridium (designed for 66+6 satellites in 6 orbital planes at 485 miles (780 km)). There are many "Little LEO" systems. "Little LEO" systems normally provide mobile data services such as Orbcomm and Teledesic. One particular example of such a system was PoSat, built by SSTL in 1993 and launched into a 510 x 497 mile (822 x 800) km orbit, inclined at 98.6 degrees.

Polar Orbit

Polar-orbiting satellites provide a more global view of Earth, circling at near-polar inclination (the angle between the equatorial plane and the satellite orbital plane -- a true polar orbit has an inclination of 90 degrees). The orbit is fixed in space, and the Earth rotates underneath. Therefore, a single satellite in a polar orbit provides in principle coverage to the entire globe, although there are long *periods* during which the satellite is out of view of a particular ground station. Most desired orbits are between 435 and 497 miles (700 and 800 km) altitude with orbit periods between 98 and 102 minutes.

This orbit provides global daily coverage of the Earth with higher resolution than geostationary orbit. Even though satellites do not pass directly over the poles they come close enough that their instruments can scan over the polar region, providing truly global coverage.

Advantages of this type of orbit are: the satellite can sample each region at the same time each day (regular schedule); high resolution due to low orbit Polar coverage (except right at the pole); one can observe the whole earth from one satellite; it is relatively inexpensive to launch into orbit (low orbit); the shadows from clouds can be used to estimate cloud heights. Some disadvantages are the poor time resolution (12 hours) and it requires a sophisticated ground system.

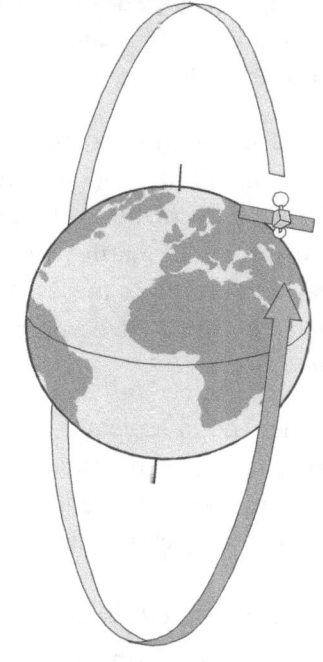

Figure 4-19: Polar Orbit

Once again, the mission of the satellite must be considered when determining the orbit type to be used. In the case of remote sensing and weather missions, the emphasis is placed high resolution and better multi-spectral coverage more than constant global coverage. Examples of satellites in polar orbits are NOAA/POES and DMSP (weather satellites) and Quickbird and Ikonos (remote-sensing satellites). Another particular example of a system that uses this type of orbit is the COSPAS-SARSAT Maritime Search and Rescue system. This system uses 8 satellites in 8 near polar orbits: Four SARSAT satellites move in 535 mile (860 km) orbits, inclined at 99 degrees, which makes them sun-synchronous. Four COSPAS satellites move in 621 mile (1000 km) orbits, inclined at 82 degrees. You will also find COSPAS-SARSAT   on Geosynchronous satellites.

Sun-Synchronous Orbit

There are situations where we want a satellite to pass over points on the Earth at the same time every day so that sunlight conditions are

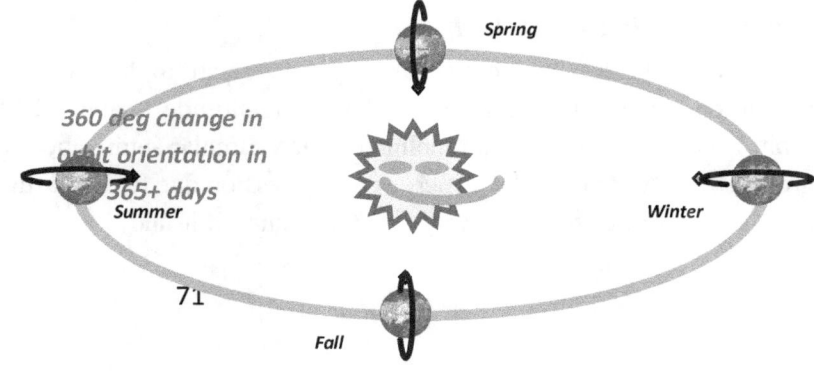

Figure 4-20: Sun Synchronous Orbit

the same. In a Sun-synchronous or Helio-synchronous orbit, the angle between the orbital plane and Sun remains constant, which results in consistent light conditions of the satellite. This can be achieved by a careful selection of orbital height, eccentricity and inclination which produces a precession of the orbit (node rotation) of approximately one degree eastward each day, equal to the apparent motion of the sun. This condition can only be achieved for a satellite in a retrograde orbit. Most commonly the inclination is about 98°.

A satellite in sun-synchronous orbit crosses the equator and each latitude at the same time each day and it will always maintain the same relative orientation to the position of the sun. This type of orbit is therefore advantageous for an earth observation satellite, as it provides constant lighting conditions. For example, the shadows cast by objects on the surface of Earth at any given latitude are always the same length when the satellite passes overhead. Thus, any change, such as new construction, is easily observed. Some weather and remote sensing environmental satellites use this type of orbit.

The sun-synchronous orbit takes advantage of a perturbation caused by the oblateness of the earth. Since the Earth is not a perfect sphere, the gravitational pull on the satellite in a polar orbit is not uniform. The result is that the orbit twists a little bit. This small twist compensates for the daily shift in position of the sun as the Earth rotates around the Sun each year. The greater gravitational force at the Equator causes the orbit to precess to the East. This precession is slightly less than one degree. This makes sense when note that the Earth rotates 360 degrees around the Sun in about 365 days.

The same sun angle every day allows photo interpreters and data analysts to rule out shadows if something in a picture has changed from one satellite pass to another. Ikonos, Geoeye, and Worldview imaging Earth resource imaging satellites are in sun-synchronous orbits for this reason. Low Earth orbiting weather satellites such as DMSP and POES polar orbiters are in sun-synchronous orbits so that weather images and data from on-board sensors are gathered at the same time every day.

Medium Earth Orbits

Medium Earth Orbits (MEO) are also called Intermediate Circular Orbits (ICO). MEOs are circular orbits at an altitude of around 10,000 km. Their orbit period measures about 6 hours. The maximum time during which a satellite in MEO orbit is above the local horizon for an observer on the earth is in the order of a few hours. A global communications system using this type of orbit requires a modest number of satellites in 2 to 3 orbital planes to achieve global coverage. MEO satellites are operated in a similar way to LEO systems. However, compared to a LEO system, hand-over is less frequent, and propagation delay and free space loss are greater. Examples of companies employing MEO orbits are ICO (10 +2 satellites in 2 inclined planes at 6422 miles (10355 km)), and Odyssey (12 + 3 satellites in 3 inclined planes, also at 6434 miles (10355 km)). Typically they are used to provide mobile telephone services.

Semi-Synchronous Orbit

A semi-synchronous orbit has a period equal to half of a day. The satellite only makes two revolutions around the Earth every day. The altitude is about 12,500 miles (20117 km) for a circular orbit. It is possible to have an inclined, nearly circular semi-synchronous orbit which repeats an identical ground trace twice each day. There is no precession due to the Earth's rotation because the 12 hour period allows the earth to turn half way around in one orbit and the rest of the way around in the second daily orbit, synching up with the same four equatorial crossing points every 24 hours. This consistent ground trace makes it easy to configure at satellite constellation to keep a certain number of satellites in view from any point on the earth.

A satellite with this period is considered to be in a medium altitude orbit. Satellites in this orbit are, however, subject to high doses of radiation in the Van Allen radiation belts. Satellites in this orbit must, therefore, be designed to withstand the increased radiation levels encountered while passing through the belts. The characteristics of this orbit make it an excellent choice for navigation systems like the US GPS satellites and the Russian GLONASS satellites.

A satellite does not necessarily have to be in a circular orbit to be semi-synchronous; they can also be highly elliptical as in the Molniya orbit that will be discussed later.

Geosynchronous Orbits

(From **geo** = Earth + **synchronous** = moving at the same rate).

A geosynchronous orbit has a period equal to that of Earth's rotation (1day). A satellite with this period is considered to be in a high altitude orbit; its mean orbital radius is about 22,300 miles. A geosynchronous orbit can have any inclination. Varying the inclination of the orbit

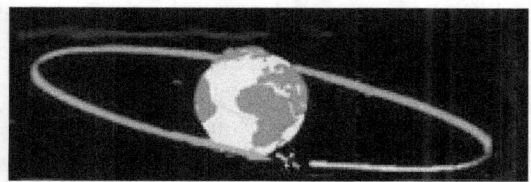

Figure 4-21: Geosynchronous Orbit

produces ground traces that fluctuate about a point on the equator in the pattern of a figure eight. The larger the inclination, the larger the figure eight, until, when at polar inclination, a figure eight ground trace with its top near the north pole and bottom near the south pole is produced.

The world's major existing telecommunications and broadcasting satellites use some type of geosynchronous orbit. Weather and surveillance/warning satellites also use geosynchronous orbits.

A satellite that appears to remain in the same position above the Earth is called a "geostationary satellite." A geostationary orbit is geosynchronous orbit with a period equal to that of the earth's rotation and an inclination equal to 0 degrees. The orbit is as circular as possible; thereby it has an eccentricity almost equivalent to zero. To an observer on the ground the satellite appears to be stationary in the sky.

The satellites can be positioned so that they are over any east/west longitude along the equator. The most significant advantage is that the satellite provides continuous coverage of specific areas of the Earth and antennas do not need to track the satellite.

The footprint, or service area of a geostationary satellite covers almost 1/3 of the Earth's surface (from about

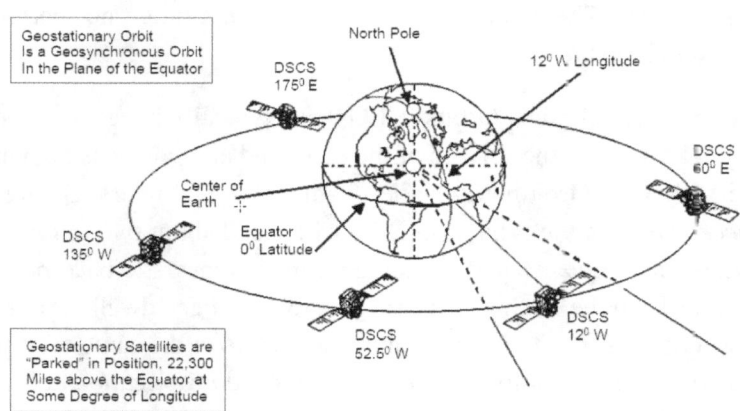

Figure 4-22: Geostationary Orbit

75 degrees South to about 75 degrees North latitude), so that a minimum of three satellites in orbit can provide global information but does not provide line of sight to the Polar Regions. The signal has to travel through much more atmosphere and the amount of interference from ground clutter makes the use of geostationary satellites impractical at latitudes greater than 70°. In the Southern Hemisphere, only Antarctica is affected. In the Northern Hemisphere, the northernmost part of Alaska along the Arctic Ocean, extreme northern Canada, Greenland, extreme northern Scandinavia and northern Siberia have restricted access to geostationary satellites nor can sensors on geostationary satellites observe these areas effectively.

Geostationary orbits are used extensively for any satellite system with a mission to provide continuous coverage or service to the same area of the world to include communications, weather, and some detection satellites. Defense Support Program satellites are in geostationary orbit so they can continually monitor an area of the Earth and provide missile warning.

Geostationary orbits are good for communications satellites like DSCS, WGS, UHF Follow On (UFO), Milstar, TV relays, and commercial communications satellites because they are always in view of the ground station and customers in the coverage area, providing continuous TV and telecommunications services to customers. Antennas on the Earth can be simple, directional antennas since they do not need to track the satellite. A disadvantage of a geostationary satellite in a voice communication system is the *round-trip delay* of approximately 250 milliseconds.

Weather satellites such as GOES are in geostationary positions to provide coverage of global weather patterns.These orbits provide a "big picture" view for coverage of weather events. This is especially useful for monitoring severe local storms and tropical cyclones. However, because a geostationary orbit must be in the equatorial plane, it provides distorted images of the Polar Regions with poor spatial resolution. The resolution measurements also suffer from the great height at which observations must be made.

Highly Elliptical Orbits

There is no universal definition for a highly elliptical orbit, however, those with an eccentricity greater than 0.5 are generally considered to be highly elliptical. A satellite in a highly elliptical orbit spends most of the time in the orbital plane toward apogee. A satellite moves slower at its apogee and fastest at its perigee. There is no specific inclination, altitude nor period associated with highly elliptical orbits.

Highly Elliptical Orbits (HEOs) can typically have a perigee at about 310 mi (500 km) above the surface of the earth and an *apogee* as high as 31,000 mi (50,000 km). Orbit *period* varies from eight to 24 hours. Owing to the high *eccentricity* of the orbit, a satellite will spend about two thirds of the orbital *period* near *apogee*, and during that time it appears to be almost stationary for an observer on the earth (this is referred to as *apogee* dwell). After this *period* a switchover needs to occur to another satellite in the same orbit in order to avoid loss of communications. *Free space loss* and *propagation delay* for this type of

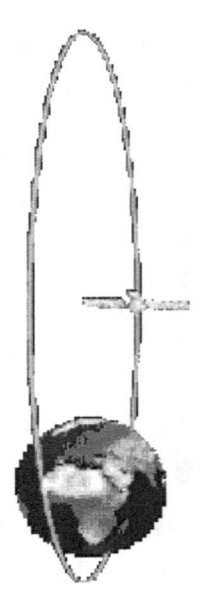

Figure 4-23: Highly Elliptical Orbit

orbit is comparable to that of geostationary satellites. The major disadvantage of the HEO orbit is that steerable antennas are normally needed at Earth terminals in order to maintain high data rate connections.

Molniya Orbit

The most common highly elliptical orbit is the Molniya orbit, originally conceived and used by the Soviet Union. This was a benefit for two reasons. First, the satellite's orbit provides better coverage of the Russian land mass for communications than a geosynchronous satellite. Communications satellites in a geostationary orbit can only provide reliable communications from about 70 degrees north latitude to 70 degrees south latitude. Much of Siberia and northern Russia is above 70 degrees north latitude. Second, the ability to launch directly into this orbit (saving on launch costs) was available simply by constructing a launch site at Plesetsk (63.4 degrees).

The orbits are inclined at 63.4 degrees (prograde) in order to provide communications services to locations at high northern latitudes. The particular *inclination* value is selected in order to avoid *rotation of the apses*, i.e. the intersection of a line from earth centre to *apogee* and the earth surface will always occur at a latitude of 63.4 degrees or 116.6 degrees North. It has an eccentricity of 0.72 and perigee over the southern hemisphere. A satellite in a Molniya orbit spends 11.7 hours of its 12 hour period in the northern hemisphere. Note that with a period of 12 hours it is also a semi-synchornous orbit. This makes the Molniya orbit well suited for communications satellites intended to provide coverage in the extreme north where access to geostationary satellite is generally not feasible. Examples of HEO systems are:

- The Russian Molniya system, which employs 3 satellites in three 12 hour orbits separated by 120 degrees around the earth, with *apogee* distance at 24,450 mi (39,354 km) and perigee at 621 mi (1000 km)

- The Russian Tundra system, which employs 2 satellites in two 24 hour orbits separated by 180 degrees around the earth, with *apogee* distance at 33320 mi (53,622 km) and perigee at 11150 mi (17,951 km)

- The proposed Loopus system, which employs 3 satellites in three 14.4 hour orbits separated by 120 degrees around the earth, with *apogee* distance at 25910 mi (41,700 km) and perigee at 3505 mi (5642 km)

OTHER ORBITAL CONSIDERATIONS

Satellite Ground Tracks

The physics of two body motion dictates the motion of the two will lie within a plane (two-dimensional motion). The orbital plane intersects the Earth's surface forming a *great circle*. A satellite's ground track is the intersection of the line between the Earth's center and the satellite and the Earth's surface;

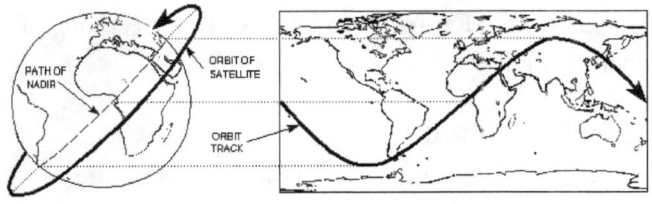

Figure 4-12: Satellite Ground Track for Single, Moderately Inclined, Circular Orbit

this point is also called the *satellite subpoint or nadir.*

The ground track (also called ground trace) of an orbiting satellite is the trace of its subpoint path across the surface of the Earth over time. A satellite ground track drawn on a map shows what areas the satellite will pass directly over. The example in **Figure 4-12** shows a moderately inclined, circular orbit and the resulting ground trace for a single orbit.

Effect of Earth's Rotation

If the Earth did not rotate, a satellite's ground track would simply repeat itself until some force changed the orbit. Of course, the Earth does rotate about its axis once (360°) each day. This equates to an angular rotation speed of about 15° per hour. The result is that each successive orbit track will be offset 15° to the west for each hour of the satellite's period. **Figure 4-13** shows the effect of the Earth's rotation for a satellite in a circular orbit with a period of 90 minutes.

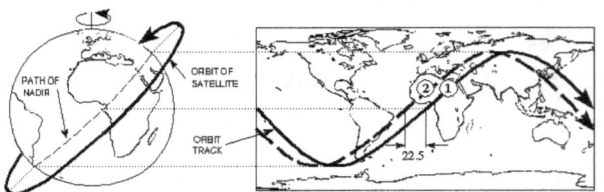

Figure 4-13: Satellite Ground Track for a Moderately Inclined, Circular Orbit (two revolutions)

Argument of perigee skews the ground track. For a prograde orbit, at perigee the satellite will be moving faster eastward than at apogee; in effect, tilting the ground track. A general rule of thumb is that if the ground track has any portion in the eastward direction, the satellite is in a prograde orbit. If the ground trace does not have a portion in the eastward direction, it is either a retrograde orbit or it could be a super-synchronous prograde orbit.

Effect of Inclination

The inclination of an orbit determines the highest north and south latitude of the ground track. Since all satellite orbital planes must pass through the center of the Earth, a satellite's ground track must be over or pass through the Equator. For a prograde orbit, the inclination of the orbit equals the highest north or south latitude of the ground track. A polar orbit has an inclination at or near 90°. The highest inclination

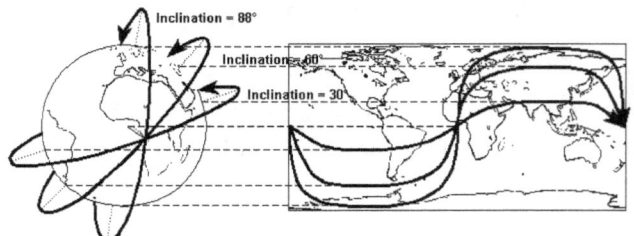

Figure 4-14: Effect of Inclination on a Satellite Orbit

of the orbit equals 180° minus the highest latitude of the nadir for a retrograde orbit (90° to 180° inclination). **Figure 4-14** shows the effect on the ground track of a satellite's orbit when only the inclination is changed.

Effect of Altitude

As the altitude of an orbit increases, the period becomes longer and the satellite's speed is lower. The result is that it takes more

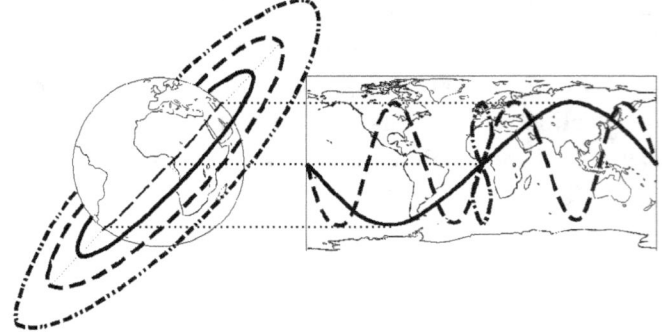

Figure 4-15: Effect of Altitude on a Satellite Orbit

complete orbits to make a complete track around the globe. Figure **4-15** shows the effect on the ground track when only the altitude is changed.

Effect of Eccentricity

Eccentricity affects the ground track because the satellite spends different amounts of time in different parts of its orbit (it's moving faster or slower). This means it will spend more time over certain parts of the Earth than others. This has the effect of creating an unsymmetrical ground track. The ground track of a satellite in a circular orbit (eccentricity = 0) is similar to a continuous sine wave. The portion above the equator is identical to the portion below. As the eccentricity of an orbit increases from zero to 1, the perigee remains close to the Earth while the apogee increases in distance. For highly eccentric orbits, the ground track of the portion near the perigee is wider than near the apogee. This is because the maximum speed of the satellite occurs at perigee and, therefore, the satellite ground track covers a greater portion of the Earth's surface. It also means at perigee a satellite is in view of a ground observer for less time. An advantage of a highly eccentric orbit is that as the satellite approaches the apogee of the orbit, it slows down and seems to linger over a geographic area. **Figure 4-16** shows the effect on a satellite's ground track caused by a change in the eccentricity of the orbit.

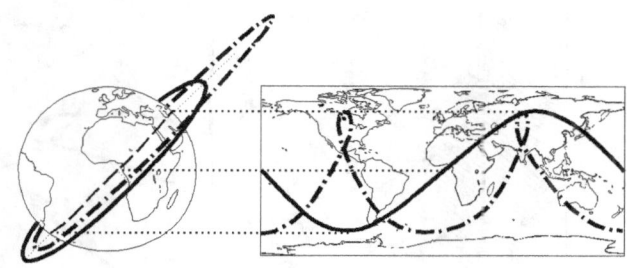

Figure 4-16: Effect of Eccentricity on a Satellite Orbit

Effect of Launch Site Location

The problem of launching satellites comes down to geometry and energy. If there were enough energy, satellites could be launched from anywhere at any time into any orbit. However, as energy is limited, the most cost efficient methods must be utilized (which usually equates to the most energy efficient). By looking at the geometry, the launch site must pass through the orbital plane to be capable of directly launching into that plane. Imagine a line drawn from the center of the Earth through the launch site and out into space. After a day, this line produces a conical configuration due to the rotation of the Earth. A satellite can be launched into any orbital plane that is tangent to, or passes through, this cone. Thus, the lowest inclination that can be achieved by directly launching is the latitude of the launch site. For launches due east (90° azimuth) or due west (270° azimuth) from a launch site, the resulting orbital inclination will equal the latitude of the launch site. After the satellite has achieved orbit, controllers must conduct additional orbital maneuvers to attain an inclination that is lower than the latitude of the launch site.

A satellite launched on an azimuth between 0° and 180° will have an inclination of between 0° and 90°, hence a prograde orbit. Satellites launched on azimuths between 180° and 360° will have an inclination between 90° and 180°, hence a retrograde orbit. Launching due north or due south (0° and 180° azimuth, respectively) will result in a polar orbit. It doesn't matter where the launch site is, the orbit will be polar.

The Earth spins to the east at 1,037 miles-per-hour (mph) at the equator and 0 mph at the poles. A substantial savings in rocket propellant is possible by locating the launch site on or near the equator and launching to the east. This gives the launcher up to 1,037 mph more velocity without burning fuel if launched due east. To launch a satellite into a retrograde orbit it is desired to launch from as high a latitude as possible so that the launcher has to overcome as little of the Earth's rotational speed as possible.

Figure 4-17 shows the allowable launch azimuths and resulting initial orbital inclinations for Cape Canaveral AFS and Kennedy Space Center in Florida.

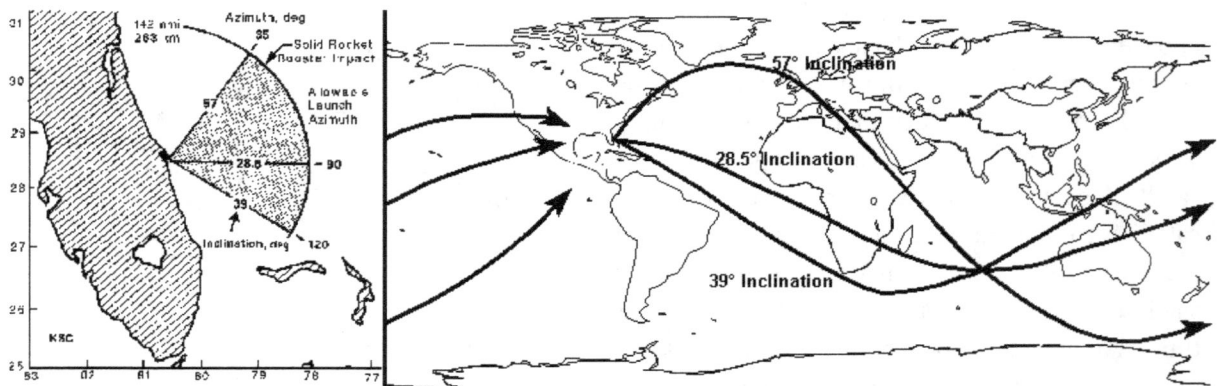

Figure 4-17: Launch Azimuth Limits and Orbital Inclinations from Kennedy Space Center, FL

Figure 4-18 shows the range of allowable launch azimuths and resulting initial orbital inclination for launches from Vandenberg AFB, CA.

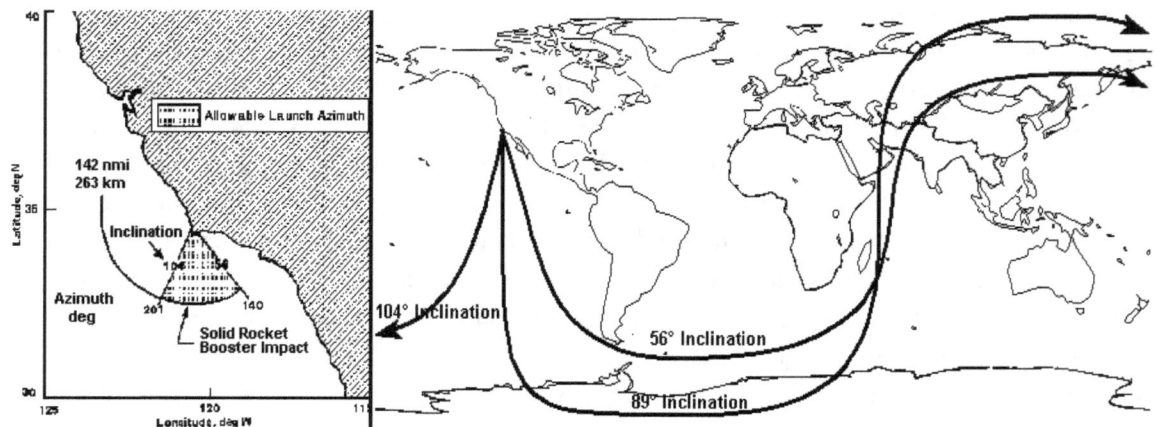

Figure 4-18: Launch Azimuth Limits and Orbital Inclinations from Vandenberg AFB, CA

Orbital Perturbations

Kepler's laws of planetary motion and Newton's laws of motion and law of universal gravitation describe how two objects move in a uniform environment. The universe, however, is not uniform. The Earth does not have a uniform density and it is not a perfect sphere. The Earth has an atmosphere and a magnetic field which extend out into space. The Sun emits vast amounts of matter and energy that vary significantly. The solar system is made up of many massive objects. The result is that there are variations in the movement and path of objects in orbit around the Earth. These variations are called perturbations. Perturbations are the effect of a variety of outside forces that act on a satellite to change its orbit.

Oblateness of the Earth

The Earth is not a perfect sphere. It is somewhat misshapen at the poles and bulges at the equator. This squashed shape is referred to as *oblateness*. The North Polar Region is more pointed than the flatter South Polar Region, producing a slight "pear" shape. The equator is not a perfect circle; it is slightly elliptical. The effect of Earth's oblateness is gravitational perturbations, and, as such, has a greater influence the closer a satellite is to the Earth. The oblateness of Earth also causes satellites in an inclined orbit to precess. Satellites in a prograde orbit precess to the west. Satellites in a retrograde orbit precess to the east. The amount of precession is more pronounced in low altitude satellites. This precession can be an advantage to help maintain a satellite in a sun synchronous orbit without expending any fuel for course changes.

Homogeneousness of the Earth

The Earth is not homogeneous. It does not have uniform density; therefore the force of gravity from the Earth's mass is not quite the same in all directions. There is a region over the Andes Mountains in South America where gravity is somewhat weaker. This is often referred to as a "gravity hill." A satellite in geostationary orbit over this region requires more station keeping because the satellite tends to drift away to other areas with stronger gravity potential. There is an area over the Indian Ocean where gravity is somewhat stronger. This is often referred to as a "gravity well". Satellites in geostationary orbit over this region tend to remain there with very little station keeping. The area over the Indian Ocean is useful for holding inactive satellites in a temporary storage location.

Atmospheric Drag

The Earth's atmosphere does not suddenly cease, rather it trails off into space. However, after about 1,000 km (620 miles), we can disregard its minuscule effects. Generally speaking, atmospheric drag can be modeled in predictions of satellite position. The highest drag occurs when the satellite is closest to the Earth (at perigee).

Third Body Effects

According to Newton's law of Universal Gravitation, every object in the universe attracts every other object in the universe. This force affects our satellites' orbits. The farther a satellite is from the Earth, the greater the third body forces are in proportion to Earth's gravitational force, and therefore, they have a greater effect on the high and deep space satellites. The greatest third body effects come from those bodies that are very massive and/or close such as the Sun, Jupiter and the Moon.

Radiation pressure

The Sun is constantly expelling atomic matter (electrons, protons, and Helium nuclei). This ionized gas moves with high velocity through interplanetary space and is known as the solar wind. The satellites are like sails in this solar wind, alternately being speeded up and slowed down, producing orbital perturbations.

Electromagnetic Drag

Satellites are continually traveling through the Earth's magnetic field. With all the electronics, the satellites produce their own localized magnetic fields that interact with the Earth's causing unwanted torque. In some instances, this torques is advantageous for stabilization. Satellites are basically a mass of conductors. Passing a conductor through a magnetic field causes a current in the conductor, producing electrical energy. Some recent experiments use a long tether in space for, among other reasons, using Earth's magnetic field to generate electrical power.

The electrical energy generated by the interaction of the satellite and the Earth's magnetic field comes from the satellite's kinetic energy about the Earth. The satellite looses orbital energy just as it does with atmospheric drag due to this energy transference. The magnetic field is strongest and satellites travel faster closer to the Earth, resulting in the largest effect on low orbiting satellites. However, the overall effect due to electromagnetic forces is quite small.

Deorbit and Decay

So far the concern has been with placing and maintaining satellites in orbit. Unfortunately, satellites are not a perpetual invention, and when no longer useful, they must be removed from their present orbit. Sometimes natural perturbations take care of disposal, but not always.

For satellites passing close to the Earth (low orbit or highly elliptical orbits) satellites can be programmed to re-enter, or they may re-enter autonomously. Deliberate re-entry of a satellite with the purpose of recovering the vehicle intact is *deorbiting*. This is usually done to recover something of value: men, experiments, film, or the vehicle itself. The natural process of spacecraft (or any debris: rocket body, payload, or piece) eventually re-entering Earth's atmosphere is *decay*.

In some situations, the satellites are in such stable orbits that natural perturbations will not do the disposal job for us. In these situations, the satellite must be removed from the desirable orbit. To return a satellite to Earth without destroying it takes a considerable amount of energy. Obviously, it is impractical to return old satellites to Earth from a high orbit. The satellite is usually boosted into a slightly higher orbit to get it out of the way, and there it will sit for thousands of years to come.

Orbital Maneuvers

A satellite is rarely launched directly into its final orbit. After being launched into an initial parking orbit, at least one orbital maneuver is needed to get the satellite into the correct orbit to begin its mission. Orbital maneuvers are necessary to maintain a satellite in its proper orbit or to change the orbital parameters based on revised mission requirements. Changing the orbital parameters of a satellite requires careful planning and precise execution. Satellites carry a limited amount of fuel to power maneuvering engines. Once the fuel has been expended there is no way to change a satellite's orbit or even to correct for small changes resulting from perturbations. There are two basic categories of orbital maneuvers: in-plane maneuvers and out-of-plane maneuvers.

In-Plane Maneuvers

In-plane maneuvers can change the size and shape of the orbital plane but do not change its orientation. In-plane maneuvers include changing the altitude of the apogee or perigee, the eccentricity of the orbit, the length of the semi-major axis, and the argument of perigee.

Hohmann Transfer Maneuver

The most energy efficient in-plane maneuver is known as the Hohmann Transfer. It is a two impulse maneuver between two coplanar orbits. The first burn is made at the perigee of the initial orbit to increase the speed of the satellite and change the eccentricity of the orbit. The magnitude and direction of the change, called the "delta-v", must be precisely controlled. If the new, more elliptical orbit is the desired final orbit, then no other burn is needed. If the final desired orbit is a circular orbit with a higher altitude than the original orbit, another burn must be made at the apogee of the transfer orbit. The second burn places the satellite in a higher, more circular orbit. The Hohmann Transfer is efficient but can take up to half of one period to execute. This technique is often used to boost communications satellites from an initial low Earth orbit into a geostationary orbit.

Fast Transfer Maneuver

Another method of changing orbits is the Fast Transfer method, used when time is a factor. This method also uses two impulses. The difference between this method and the Hohmann Transfer is that the satellite approaches its new orbit at a higher angle and velocity. Thus, the second burn must be stronger to brake the satellite into the new orbit. Fast transfer is used by various surveillance satellites when new targeting requirements are critical.

Out-Of-Plane Maneuvers

Out-of-plane maneuvers change the orientation of a satellite's orbital plane. As with a spinning gyroscope, making a change to the orbital plane of a satellite can require large amounts of energy. Two common out-of-plane maneuvers are changes to the inclination and to the right ascension of the ascending node. To change only the inclination of a satellite's orbit, one burn is made at either the ascending node or the descending node. To change only the right ascension of the ascending node requires one burn anywhere in the original orbit except at the ascending or descending nodes.

SATELLITE ORBIT REFERENCES

Websites:

http://marine.rutgers.edu/mrs/education/class/paul/orbits2.html - Good site describing several orbits and providing graphical descriptions of those orbits.

http://lectureonline.cl.msu.edu/~mmp/kap7/orbiter/orbit.htm - With this applet you can put your own satellite into orbit around Earth. Keep the mouse down and drag to adjust the velocity vector. With a little practice, you can keep your own Sputnik on an orbit without it crashing into Earth or escaping.

http://www.n2yo.com - Real time satellite tracking.

http://heavens-above.com – Another satellite tracking site plus information relative to the sky charts.

Chapter 5
Space Mission Areas

OVERVIEW

A doctrinal construct used to organize the many resulting space issues is that of space mission areas. Consistent with JP 3-14, the DoD divides the space mission into four primary mission areas: space control, space force enhancement, space force application, and space support. Space capabilities are seamlessly integrated with warfighting capabilities so there is no loss of information at the interfaces, and just as important, no loss of time. Continuity is maintained among sensors, shooters, and command and control through thoughtful application of space capabilities. These four mission areas are the operational foundation of DoD space capabilities.

SPACE CONTROL

Space control operations provide freedom of action in space for friendly forces while, when directed, denying it to an adversary, and include the broad aspect of protection of US and US allied space systems and negation of adversary space systems. To gain space superiority, space forces must surveil space and terrestrial AOIs that could impact space activities, protect the ability to use space, prevent adversaries from exploiting US, allied, or neutral space services, and negate the ability of adversaries to exploit space capabilities. Space control includes offensive and defensive operations by friendly forces to gain and maintain space superiority and situational awareness of events that impact space operations. Space control involves five interrelated objectives:

- Surveillance of space to be aware of the presence of space assets and understand real-time satellite mission operations.

- Protect US and friendly space systems from hostile actions.

- Prevent unauthorized access to, and exploitation of, space systems.

- Negate hostile space systems that place US interests at risk.

- Directly support battle management, command, control, communications, and intelligence.

Surveillance

The Space Surveillance Network (SSN) mission is to detect, track, identify, catalog and characterize all man-made objects in space. The Space Control Center (SCC) operated by the 14th Air Force at Vandenberg AFB maintains a catalog on objects down to 10 cm. in diameter, approximately the size of a softball, for near earth. Objects are maintained at geosynchronous ranges to 100cm. in diameter, approximately the size of a large beach ball. The surveillance system is used to:

- Predict when and where a decaying space object will re-enter the Earth's atmosphere;

- Prevent a returning space object, which might resemble a missile, from triggering a false alarm in missile-attack warning sensors of the US and other countries;

- Chart the present position of space objects and plot their anticipated orbital paths;

- Detect satellite maneuvers;

- Detect satellite breakups;

- Detect new man-made objects in space;

- Produce a running catalog of man-made space objects;

- Determine which country owns a re-entering space object;

- Inform NASA whether or not objects may interfere with International Space Station orbits.

Because of the limits of the SSN, which include the number of sensors, geographic distribution, capability and availability, every satellite cannot be continuously tracked. To maintain a database of all man-made objects in earth orbit, the SCC uses a tracking cycle, which starts with a prediction. The SCC makes an assumption as to where a newly launched object will be, and then sends out this postulation in the form of an element set (ELSET) to the sensors. Subsequently, the sensor uses this ELSET to search for the object. If the assumption is close, the sensor will detect-and-track the object. The sensor then collects observations, which are sent back for processing and analysis.

The SCC uses this information to compute a new ELSET, or prediction, which is then sent to the next sensor to track or observe it. This cycle repeats 24 hours a day, seven days a week.

All sensors in the SSN are responsible for providing space surveillance and Space Object Identification (SOI) to the SCC. The sensors in the network are categorized primarily by their availability for support to the SCC. There are four main sensor categories: dedicated (active), collateral, contributing and passive.

Dedicated Sensors

A dedicated sensor is an operationally controlled sensor with a primary mission of space surveillance support. Dedicated sensors include GEODSS, MOTIF, the Space Surveillance Fence, and AN/FPS 85 PAR.

GEODSS, or Ground Based Electro-Optical Deep Space Surveillance System, has the mission to detect, track and collect SOI on deep space satellites. GEODSS sites are controlled and operated by the 21st Space Wing at Peterson AFB, Colorado. The GEODSS sites provide near real-time deep space surveillance capability. To perform its mission, GEODSS brings together the telescope, low-light television and computers. Each site has three telescopes; two main and one auxiliary (with the exception of Diego Garcia, which has three main telescopes). The system only operates at night, when the telescopes are able to detect objects 10,000 times dimmer than the human eye can detect. Since it is an optical system, cloud cover and local weather conditions influence its effectiveness. The GEODSS system can track objects as small as a basketball, more than 20,000 miles in space.

MOTIF stands for Maui Optical Tracking and Identification Facility. MOTIF performs near-earth/deep space surveillance and SOI. It uses not only photometric and visual imaging, but also a long wave infrared (LWIR) data generation system. It is an optical sensor very similar to the GEODSS sites, in addition to its LWIR capability. MOTIF consists of a dual telescope system on a single mount. One telescope is used primarily for infrared and photometric (light intensity) measurements. The other is used for low-light-level tracking and imagery. Both use video cameras to record data. MOTIF has identified objects as small as 8 cm in geosynchronous orbit.

The Space Surveillance Fence is the oldest sensor system in the SSN. Originally know as Navy Fence, the 20th Space Control Squadron through Detachment 1 at Dahlgren Naval Base manages the fence mission to maintain a constant surveillance of space and to provide satellite data. The Space Surveillance Fence uses three transmit antennas and six receive antennas, all geographically located along the 33rd parallel of the US The transmitters send out a continuous wave of energy into space, forming a "detection fence," which covers 10 percent of the Earth's circumference and extends 15,000 miles into space. When a satellite passes through the fence, the energy from the transmitter sites "illuminates" it and a portion of the energy is reflected back to a receive station. When the reflected energy is acquired by at least two receive sites, an accurate position of the satellite can be determined through triangulation.

The AN/FPS 85 PAR located at Eglin AFB, Florida and operated by the 20th Space Control Squadron, is the only phased array radar (PAR) space surveillance system dedicated to tracking space objects. It is one of the earliest PARs, built in the mid-1960s and became operational in December 1968. It has the capability to track both near-earth and deep space objects simultaneously. The previous primary mission at Eglin was Submarine- Launched Ballistic Missile (SLBM) warning. Once PAVE PAWS South East (at Robins AFB) became operational, the SLBM warning coverage was redundant and Eglin's mission changed in 1988 to dedicated space surveillance. The 20th Space Control Squadron tracks over 95 percent of all earth satellites daily.

The Space Based Space Surveillance (SBSS) spacecraft launched in September 2010 detects deep space objects and features a digital optical sensor mounted on a high-speed, two-axis gimbal. That allows ground controllers to quickly swivel the camera between targets without having to expend the time and fuel to reposition the entire spacecraft. This agile sensor mount enables SBSS to find and track objects in space – even new spacecraft launches and maneuvers – with significantly greater speed, capacity and sensitivity than previous space sensors. In orbit, SBSS sensors are not affected by weather, atmosphere or time of day, and it has a much wider field of view than sensors on the ground.

Collateral Sensors

A collateral sensor is a USSTRATCOM operationally controlled sensor with a primary mission other than space surveillance (usually, the site's secondary mission is to provide surveillance support). Collateral sensors include BMEWS, PAVE PAWS, PARCS, Antigua, and Ascension.

Ballistic Missile Early Warning System (BMEWS) is a key radar system developed to provide warning and attack assessment of an Intercontinental Ballistic Missile (ICBM) attack on the CONUS and southern Canada from the Sino-Soviet landmass. BMEWS also provides warning and attack assessment of a SLBM/ICBM attack against the United Kingdom and Europe. BMEWS' tertiary mission is to conduct satellite tracking as collateral sensors in the space surveillance network. BMEWS consists of three sites: Site I is located at Thule AFB, Greenland; Site II is at Clear AFB, Alaska; and Site III is at Royal Air Force Station Fylingdales, United Kingdom.

The PAVE PAWS is a dual-faced, phased array radar (AN/FPS-115) and provides warning and attack assessment of an SLBM attack against the CONUS and southern Canada. PAVE PAWS also provides warning and attack assessment of an ICBM attack against North America from the Sino-Soviet land mass. The final tertiary mission, like BMEWS, is to provide satellite tracking data as collateral sensors in the space surveillance network. PAVE PAWS currently consists of the initial two sites: Site I is located at Cape Cod AFS, Massachusetts; Site II is at Beale AFB, California.

Perimeter Acquisition Radar Attack Characterization System (PARCS), originally built as part of the Army's Safeguard antiballistic missile (ABM) system, is located just 15 miles south of the Canadian border at Cavalier AS, North Dakota. It was operational for one day (1 October 1975). It now provides warning and attack characterization of an SLBM and ICBM attack against the CONUS and southern Canada. It is one of the workhorses, along with Eglin, providing surveillance, tracking, reporting and SOI data.

The radars located on the Antigua and Ascension Islands in the Atlantic Ocean are tracking radars. They are part of the Eastern Range, playing a secondary role in supporting the SCC.

Contributing Sensors

The contributing sensors are those owned and operated by other agencies, but provide space surveillance upon request from the SCC. They include Millstone/Haystack, ALTAIR and ALCOR, Kaena Point, and AMOS.

The Millstone Hill Radar and Haystack Long Range Imaging Radar in Lexington, Massachusetts is owned and operated by Lincoln Laboratories of the Massachusetts Institute of Technology (MIT). Millstone is a deep space radar that contributes 80 hours per week to the SCC. Haystack is a deep space imaging radar that provides wideband SOI tracks to the SCC. Haystack provides this data one week of every six of Millstone operations.

The Advanced Research Project Agency (ARPA) Long-Range Tracking and Identification Radar (ALTAIR) and the ARPA Lincoln C-Band Observable Radar (ALCOR) are located in the Kwajalein Atoll in the western Pacific Ocean. Operated by the Army, they are primarily used for ABM testing in support of the Western Range. They support the space surveillance mission when possible. ALCOR is a near-earth tracking radar, and is the only other radar besides Haystack, that can provide wideband SOI. ALTAIR is a near-earth and deep space tracking radar. Because of its proximity to the equator, ALTAIR alone can track one-third of the geosynchronous belt.

Kaena Point is a tracking radar located on Oahu, Hawaii. It is part of the Western Range and supports the SCC with satellite tracking data.

The last contributing sensor is the ARPA Maui Optical Station (AMOS), which is a GEODSS-type optical sensor collocated with the GEODSS and MOTIF sensors on Maui.

Passive Space Surveillance System

The US passive space surveillance system consists of four worldwide units that locate and track man-made objects in space via a new generation of radio frequency technology. There are two types of passive space surveillance units: the Deep Space Tracking System, or DSTS, and the Low-Altitude Space Surveillance System, or LASS.

Deep space tracking involves tracking objects with orbits that take more than 225 minutes to rotate the earth (geosynchronous). DSTS antennas are at the Misawa AB, Japan and Feltwell, U.K. Low altitude satellites have orbits less than 225 minutes. LASS systems are also at Feltwell, U.K and Osan AB, Republic of Korea.

Protection

Active and passive defensive measures ensure that US and friendly space systems perform as designed by overcoming an adversary's attempts to negate friendly exploitation of space or minimize adverse effects if negation is attempted. Such measures also provide some protection from space environmental factors. Protection measures must be consistent with the criticality of the mission's contribution to the warfighter and are applied to each component of the space system, including launch, to ensure that no weak link exists. Means of protection include, but are not limited to, ground facility protection (security; covert facilities; camouflage, concealment, and deception; mobility), alternate nodes, spare satellites, link encryption, increased signal strength, adaptable waveforms, satellite radiation hardening and space debris protection measures. Furthermore, the system of protection measures should provide unambiguous indications of whether a satellite was under attack or in a severe space weather environment when any satellite anomaly or failure occurs. Finally, attack indications could be so subtle or dispersed that individually, an attack is not detectable.

Prevention

Prevention includes measures to preclude an adversary's hostile use of US or third party space systems and services. Prevention can include military, diplomatic, political, and economic measures as appropriate. An example of this might be the encryption of telemetry and commanding signals to satellites or even the encryption of GPS receivers.

Negation

Measures to deceive, disrupt, deny, degrade, or destroy an adversary's space capabilities. Negation can include action against the ground, link, or space segments of an adversary's space system.

- Deception includes measures designed to mislead the adversary by manipulation, distortion, or falsification of evidence to induce the adversary to react in a manner prejudicial to their interests

- Disruption is the temporary impairment (diminished value or strength) of the utility of space systems, usually without physical damage to the space system. These operations include the delaying of critical, perishable operational data to an adversary.

- Denial is temporary elimination (total removal) of the utility of an adversary's space systems, usually without physical damage. This objective can be accomplished by such measures as interrupting electrical power to the space ground nodes or computer centers where data and information are processed and stored. For example, denying US adversaries position navigation information could significantly inhibit their operations.

- Degradation is the permanent partial or total impairment of the utility of space systems, usually with physical damage. This option includes attacking the ground, control, or space segment of any targeted space system. All military options, including special operations, conventional warfare, and information warfare are available for use against space targets.

- Destruction is permanent elimination of the utility of space systems. This last option includes attack of critical ground nodes; destruction of uplink and downlink facilities, electrical power stations, and telecommunications facilities; and attacks against mobile space elements and on-orbit space assets.

THREAT TO SPACE SYSTEMS

Overview

The nation's dependence on space capabilities in the 21st century rivals its dependence on electricity and oil in the 19th and 20th century. Military operations depend critically on space capabilities. In the 21st century they will rely even more on such services as global communications, reconnaissance and surveillance in near real time, accurate missile warning, and position/navigation. Space capabilities are critical for integrating the effects of widely dispersed weapons platforms and forces.

In the civilian sector, space has become even more critical to industry and international business operations. As the US economy moves from an industrial-focused nation to an information-focused nation, space becomes a vital national interest. The recent National Security Strategy and National Space Policy have indicated that freedom of action in space is a US interest. Space has or will soon emerge as a center of gravity for DoD and the nation. Space and Information Dominance have already become key tenants for future warfare. We must commit enough planning and resources to protect and enhance our access to, and use of, space.

Various sources have addressed the threat to ensuring our freedom of action in space. Recent activity, such as the kinetic satellite destruction by China in 2007, suggests the sanctity and freedom of operation in Space is tenable at best. Most of the information addressing the actual threat can only be found in classified sources. We need to have some understanding of what is the future strategic environment that will affect space interests, some current threat offensive capabilities and the offensive measures we may have available. When discussing threats to space it is important to remember that space systems include both satellites and ground facilities along with command, data, and communications links between them.

FUTURE STRATEGIC ENVIRONMENT

There are a number of assumptions or key judgments that are made about the future strategic environment that will influence how wars may begin and be fought. The assumptions listed below are just a few that relate specifically to US space interests. Threats to our national interests will continue to emanate from these sources.

- Rapid advancement of technology will create revolutionary breakthroughs. Commercial interests will drive most technology development, especially with space and information processing.

- More nations and non-state actors will access space and information technologies and products. Rapid integration of information enabled by space capabilities will be the key to successful operations.

- Advanced warheads and stealth technology will become more common in ballistic missiles.

- The precision and lethality of future weapons, to include space-based weapons, will lead to increase massing of effects rather than massing of forces.

- Technology will proliferate for weapons of mass destruction and delivery systems.

- Space capabilities will proliferate at a faster pace.

- The shift will continue from the military to the commercial sector as the dominant receiver and provider of space services.

- Military, civil and commercial space sectors will continue to converge.

- Achieving space superiority or denying space to the adversary during conflicts will be critical to any military's success.

- Sovereignty will remain a concern and space "basing rights" will become a bigger issue.

- Space capabilities will be increasingly important to our society and economy and they will become vital US national interests.

THREAT CHARACTERISTICS

Threats will be tactically and technologically diverse. Armed conflict will remain a vicious, lethal business. Opposing forces may appear in regular or paramilitary structures, transnational military organizations, and single or widespread domestic and international terrorist cells. Criminal elements of various sizes and means can probe our geographical, human, electro-magnetic, and cyber borders. They may threaten US interests abroad or populations, critical infrastructure, or territory at home. Some adversaries will enhance their C4ISR with international or commercial space systems and the global information infrastructure. Adversaries will wield technologically advanced niche capabilities, and develop asymmetric capabilities for waging computer network and information warfare. Some may employ precision fires or massed fires to put our forces or partners at risk during mobilization, deployment, and operations or redeployment. The threat of nuclear, chemical and biological warfare will persist, as will the specter of attack with strategic or intercontinental ballistic and cruise missiles.

Many adversaries will be politically astute and innovative. Adversaries will attempt to undermine the national will to conduct operations and fracture the cohesion of coalitions and alliances. This approach is now enabled by the worldwide proliferation of telecommunications and information technology. Traditional foreign intelligence collection against the US will remain a concern. However, the effort will evolve in new directions, stemming from reliance on computer systems for processing and storing sensitive information. Space systems allow the US to monitor and report on global activities, observe early indications of crises, provide planning information, reduce many uncertainties characterizing conflict situations, and support information sharing with partners and allies. Space operations can do much to preclude such threats, but our adversaries can put space systems at risk.

Adversaries may put at risk all elements of space systems. Adversaries may alter the space environment, on-orbit assets, communications links, ground stations, terminals, or the associated information infrastructure. A variety of direct ascent and co-orbital anti-satellite (ASAT) techniques will be possible. On the ground, aerospace industrial facilities and launch sites may come under attack, as well as boosters during ascent. If vulnerable, ground stations, control facilities, and terminals will be attacked by enemy special forces, terrorists, theater missiles, electronic warfare means, cyber-attack, or conventional forces. Electronic attacks will aim to degrade satellite communications; telemetry, tracking, and control (TT&C) links; and ground stations. Low power signals such as GPS are particularly susceptible to localized interference.

Space systems will be in the IO target set, and the linkage between information operations and space operations has been forged by the assignment of selected IO responsibilities to US Strategic Command. Commander, USSTRATCOM, must prepare to counter computer hacking aimed at tampering with streams of data moving to and from satellites. If vulnerable, space systems will come under computer network attack (CNA). Left unprotected, links will be jammed, monitored, or pirated by adversaries.

Adversaries will also use space. Commercial and international space is expanding our capabilities, as well as our adversaries'. Twelve hundred new commercial satellites will be on orbit in the next decade. The majority will be communications systems, but thirty or more imaging satellites are likely to be operating. As a result, states, transnational organizations, factions, or individuals will be able to buy militarily significant space products or services. In fact, 1-meter resolution imagery, sufficient for tactical targeting (if timely) is already internationally available. Other purchases will include radar imagery, precision navigation and timing services, and a multitude of highly mobile, highly capable communications. While licensing and international pressure will impose some restrictions, this burgeoning space industry has been termed the new El Dorado. The lure of a space gold mine, coupled with the rise of the global economy, will dampen voluntary controls.

The systems are not all commercial, however, as more than 40 nation-states have national space programs of varying sophistication. So, military operations must assume an adversary will have limited access to overhead intelligence capabilities and telecommunications capable of operating in remote, undeveloped areas, as well as in urban environments.

Offensive Counterspace Measures

Advanced technology will help develop counterspace weapons aimed against US or allied space systems. Although improved satellite systems will depend less on ground infrastructures for TT&C, they will still be susceptible to directed-energy weapons that can permanently or temporarily disable critical satellite functions. Keep in mind that space systems consist of space (mission and data relay satellites), ground (supporting ground facilities), and user segments. The sophisticated enemy's hostile capabilities may include kinetic, electronic, and direct-energy systems to negate US satellites. A lesser nation state or other non-state actors may prefer jamming or attacks against ground and user systems. Deception and information operations against our C4I systems may also be use.

Offensive counterspace operations neutralize space systems or the information they provide through attacks on the various elements of those systems. They involve the use of lethal or non-lethal means for negation. Deception, disruption, denial, degradation, or destruction of space systems and services may seriously affect the US commercial, civil and military environment. Most countries already have the capability (but not necessarily the will) to affect the space environment. There are four types of offensive counterspace operations that could be used.

Denial and Deception

These techniques are used to limit or corrupt information obtained by an intelligence collection satellite. This is a highly attractive counterspace alternative because of the low level of difficulty. Denial and deception involves the employment of camouflage, concealment, and deception techniques. Many countries have denial and deception practices to impede the collection of intelligence by an adversary. A significant advantage of denial and deception is that it can be conducted in peacetime.

In order to conduct effective denial and deception operations, one must know when intelligence satellites will be overhead. Such information can be obtained in many ways. Amateur observers post what they believe to be intelligence collection satellite data on several Internet sites. Countries more acutely concerned about satellite overflights can implement their own observing program. The accuracy of publicly available software to predict satellite orbits means some actors can use satellite warning as part of a plan for denial and deception.

Ground Segment Attack/Sabotage

Physical attacks and/or sabotage can be used against critical satellite ground facilities such as control stations or data processing stations to disrupt, deny, degrade, or destroy the space system. The greatest advantage of this method is that it can be accomplished using existing military systems. Most potential adversaries could attack the ground and user facilities associated with a space system. Critical ground facilities include TT&C facilities, launch facilities and assembly plants, satellite assembly plants, and key data processing or data downlink facilities. There can be mobile ground facilities as well as fixed facilities. Most fixed satellite ground facilities are described in open source materials and are therefore easy to identify and target.

Electronic Attack

Electronic warfare is not a new capability to many countries and can be used to disrupt or deny such things as satellite communications, position/navigation/timing data, and command links. All satellites use some form of communication link to pass information. These electronic signals can be neutralized by jamming or spoofing the electronic equipment on the satellites or at the receiving ground facility. Jammers usually emit noise-like signals in an effort to mask or prevent the reception of desired signals, while spoofers emit false but plausible signals that are received and processed along with desired signals.

It is relatively easy to jam signal uplinks and downlinks. The jammer must operate in the same radio band as the signal to be jammed. However, transmit power becomes an issue in determining how easy it is to jam the uplink or the downlink. In general, uplink jammers must be roughly as powerful as the emitter associated with the link being jammed. Since downlink jammers have a significant range advantage over the space-based emitter with which they are competing, they can often be much less powerful and still be effective.

Targets of downlink jammers are ground-based satellite data receivers ranging from large sites to handheld GPS user sets. Systems using low power receive (downlink) signals, such as GPS and some communication systems, are relatively easy to jam and are jammed many times through the use of other RF transmitters in the area. Jamming is normally local (from tens to hundreds of miles).

Targets of uplink jammers are the satellites' radio receivers, including their sensors. Uplink jamming is more difficult since considerable jammer transmit power is required. However, if effective, it can deny information globally. If false commands can be sent to the satellite, it could cause the spacecraft to destroy itself. Vulnerability of satellite systems to jamming depends strongly on particular designs.

Two areas of great concern to the military are communications and GPS. The military has become more reliant on commercial communications that are not hardened against jamming. During OIF and OEF more than 80% of all communications passed between the US and the Persian Gulf were over commercial satellite systems and could have been jammed with existing communications equipment

Space Segment Attack (Anti-satellite)

Direct attack against a satellite has historically been a capability only the US and former Soviet Union possessed. China may also have the capability to deploy anti-satellite (ASAT) weapons and has demonstrated a direct ascent kinetic kill capability in 2007. However, global technology will lead to the proliferation of ASAT threats especially as threats against LEO systems.

ASATs can be grouped into two generic categories, interceptors and directed-energy weapons. The ASAT interceptor category consists of low-altitude direct-ascent interceptors; low-altitude co-orbital interceptors; high-altitude, short duration interceptors; and long-duration orbital interceptors. Directed-energy ASAT weapons consist of ground-based high-power lasers; low-power anti-sensor lasers; airborne high-power lasers; space-based lasers; and RF weapons.

Vulnerability of satellite systems to component damage from directed energy weapons depends on the design and orbit of the satellite. LEO satellites are close to the earth allowing an enemy to concentrate large amounts of energy on them. Since imagery optical systems must concentrate incident light on a small sensitive detector, laser attacks against these sensors can be highly effective.

Enemy nations that can track satellites and fire significant payloads into space can place important satellites at risk using inexpensive direct-ascent weapons. But this ability will belong only to the most advance countries because of the cost of research, development, and implementation. Russia had previously tested a co-orbital interceptor and had developed a concept for space-to-space missile platforms. Other countries have boosters that could be used for direct-ascent ASAT weapons but not necessarily the ASAT technology.

High Altitude Nuclear Device

A high-altitude nuclear detonation would create electromagnetic interference against satellite communications. Satellites in LEO or MEO orbits would be most affected (such as Iridium, DMSP, and GPS). Prompt nuclear effects include electronic upset, electronic burnout, and mechanical damage. Delayed nuclear effects include the same effects due to trapped radiation. The detonation of a single nuclear weapon above the atmosphere could have severe, long-term consequences. One should note that this would be in direct violation of the Outerspace Treaty of 1967.

Only a few nations have the capability to launch a nuclear weapon and most nation-states wouldn't be willing to accept fratricide of their own satellite systems. However, some nation-states that don't rely heavily on space capabilities may be willing to reduce US military and industrial capabilities at the risk of nuclear fratricide.

Information Operations

Offensive information operations are cheap relative to the cost of developing, maintaining, and using direct energy or nuclear devices. Many information operations tools are readily accessible to third world nations or non-state actors. Information attacks could consist of creating false information, manipulating information, and inserting malicious, logic-based weapons in space-based, globally shared infrastructures for telecommunications and computing.

US CAPABILITIES

Control of Space is the ability to ensure uninterrupted access to space for US forces and our allies, freedom of operations within the space medium and an ability to deny others the use of space, if required. We must protect our space assets and be able to deny adversaries from gaining an advantage through their space systems. The ability to gain and maintain space superiority will become critical to military operations.

This section will discuss current ways the US is looking at negation systems. Negation will be executed when prevention fails. Negation is the ability to deny, disrupt, deceive, degrade, or destroy an adversary's space systems or services. Actions we could take range from temporarily disrupting or denying hostile space systems to degrading or destroying them. Jamming, attacks against ground segment systems and information operations are always applicable as negation tools. When considering the options, the commander must consider third-party use of a satellite, plausible deniability and how actions may add to debris or otherwise affect the environment. Some of the technologies being explored are listed below.

The space-based systems that the US is looking at are for space control and force application. All are new technology systems and there is not much information about them in open source materials.

Directed Energy Weapons

Directed energy weapons include high-energy laser and high-power microwave.

Radio Frequency (RF) Weapons

High Power Microwaves may be able to disrupt, degrade, and destroy electronics in communication and information systems. They would use bandwidths at high peak power to damage electronic information processing and communication or bandwidths at high average power to disrupt systems. These RF weapons would be deployed in GEO orbits and use large antennas to direct RF energy against enemy electronic systems.

Laser Weapons

Space based lasers appear to be the most versatile in providing for options for temporary effects on anticipated targets. The potential for using lasers has been recognized since the technology was discovered in 1960 but producing beams with enough power has been problematic. The type of laser currently being developed is the chemical laser.

Although Congress cut the Department of Defense's Space-Based Laser (SBL) acquisition program in 2000, the Missile Defense Agency (MDA) maintains an active interest in developing the technology necessary for space-based missile defense, and work within the Directed Energy Directorate of the Air Force Research Laboratory continues at the basic and applied research levels. SBLs present unique challenges for laser developers. The requirements of autonomous operation, minimal logistics, low weight, and radiation-hardened materials are very significant basic research and engineering issues. To date, only chemical lasers appear to have the potential to generate the requisite power levels and beam qualities to meet mission requirements. Furthermore, researchers have been able to scale only the chemical oxygen iodine laser (COIL) and hydrogen fluoride (HF) chemical lasers to very-high-energy systems.

The Army manages HELSTF (High Energy Laser System Test Facility) as the DoD National Test Facility for High Power Lasers. HELSTF is located at White Sands Missile Range and is capable of testing lasers over a broad range of wavelengths. HELSTF is the home of the Mid Infrared Advanced Chemical Laser (MIRACL), the United States' most powerful laser.

Direct Impact Weapons

These are weapons that use either kinetic energy (KE) or ones that pass close enough to a satellite for an exploding fragmentation device to destroy it.

KE-ASAT

According to The Boeing Company, the Kinetic Energy Anti-Satellite (KE-ASAT) program was intended to provide the United States with the capability to neutralize hostile satellites. The objective of the KE-ASAT program was to define, develop, integrate and test the necessary Kill Vehicle (KV), weapon control subsystem component and subsystems technologies to demonstrate hit-to-kill performance, including debris mitigation against hostile satellites. This technology still has application to future systems.

Development of this technology into deployable offensive or defensive systems will still take a number of years. However, some of the systems the United States is currently developing to intercept ballistic missiles would have considerable inherent capability to be used as ASAT weapons. Current-generation satellites are not equipped to defend themselves. While future satellites might include defenses of some type, it will be difficult to overcome the advantages inherent to an attacker.

Co-Orbital ASAT

A co-orbital ASAT is one that closes slowly with its target. It uses an exploding warhead to destroy the satellite. The feasibility of a co-orbital ASAT weapon was demonstrated successfully by Russia in 1982.

FORCE APPLICATION

Joint Publication 3-14, Joint Doctrine for Space Operations, states, "The application of force would consist of attacks against terrestrial-based targets carried out by military weapons systems operating in or through space." Army FM 3-14, Space Support to Army Operations, uses the basic JP 3-14 definition that states space force application is "Combat operations in, through, and from space to influence the course and outcome of conflict. The space force application mission area includes ballistic missile defense and force projection." Currently, there are no force application assets operating in space.

Ballistic Missile Defense doctrine is beyond the scope of this manual and is not addressed here. Other than BMD, the Army currently provides no space force application weapons systems. This is a potential mission of the future when space-based platforms can add an accurate and immediate third-dimensional sensor and shooter capability.

SPACE SUPPORT

Space support includes operations to deploy, sustain, and modernize military systems in space. This mission area includes launching and deploying space vehicles, maintaining and sustaining spacecraft on orbit, and deorbiting and recovering space vehicles, if required. It involves actions to sustain and maintain US space-based constellations, such as space lift, surveillance of systems in space, and the day-to-day telemetry tracking and control (TT&C) needed for optimal performance and health of assets. It is a combat service support operation to deploy and sustain military and intelligence systems in space.

Overview

Space systems can be divided into two categories: the launch vehicle and the spacecraft. The launch vehicle, commonly called the booster, propels the spacecraft and its associated payload(s) into space. Typically for military missions, a specific spacecraft is flown on a specific booster. For example, the Global Positioning System (GPS) satellites are always launched on a Delta II launch vehicle.

Missiles and space have a long and related history. All the early space boosters, both US and Russian, were developed from ballistic missile programs. Today, many other nations are using their missile and rocket technology to develop a space launch capability.

Launch Sites

There are at least 24 launch sites world wide including Sea Launch. Many of these sites are not commonly used or are used for sounding rocket launches only. Some of the sites are still under development. This section will discuss a few of the sites shown in **Figure 6-11**. The below list is not all-inclusive.

1 – Vandenberg AFB	9 – Andova	17 – Xichang
2 – Edwards AFB	10 – Plesetsk	18 – Taiyuan/Wuzhai
3 – Wallops Island	11 – Kapustin Yar	19 – Svobodny
4 – Cape Canveral/KSC	12 – Palmachim/Yavne	20 – Kagoshima
5 – Kourou	13 – San Marco Platform	21 - Tanegashima
6 – Alcantara	14 – Baikonour/Tyuratam	22 - Woomera
7 – Hammaguir	15 – Sriharikota (SHAR)	23 – Sea Launch
8 – Torrjon AB	16 – Jiuquan	24 – Kodiak Island, AK

US Launch Facilities

Eastern Range

The Eastern Range is located on the east coast of Florida (28.5° N, 80° W). The ESMC includes Cape Canaveral Air Force Station where most of the launch pads are located, Patrick AFB where the headquarters is, the Eastern Test Range and other supporting facilities in east central Florida. The Eastern Test Range (ETR) extends from Cape Canaveral, across the Atlantic Ocean and Africa into the Indian Ocean. The ETR includes tracking stations on Antigua and Ascension Island.

The United States' largest space launch facility is located at Cape Canaveral Air Force Station. Since 1950, more than 40 launch complexes have been constructed. Some of the launch pads were built to test ICBMs and Submarine Launched Ballistic Missiles. Current launch vehicles include Delta, Atlas, and Athena. With the cancellation of the Shuttle Transport System, the U.S. is looking to promote commercial launch capabilities and companies like SpaceX are launching from this range as well.

Cape Canaveral is located at 28.5° north latitude. The optimum launch (most fuel efficient or heaviest payload) is attained by launching directly to the east (azimuth of 90°) to take maximum advantage of the Earth's rotational speed, thus the minimum inclination of a satellite's initial orbit is 28.5°. Safety considerations limit the launch azimuth to a minimum of 35° to a maximum of 120° for an initial orbit inclination of 57° and 39°, respectively.

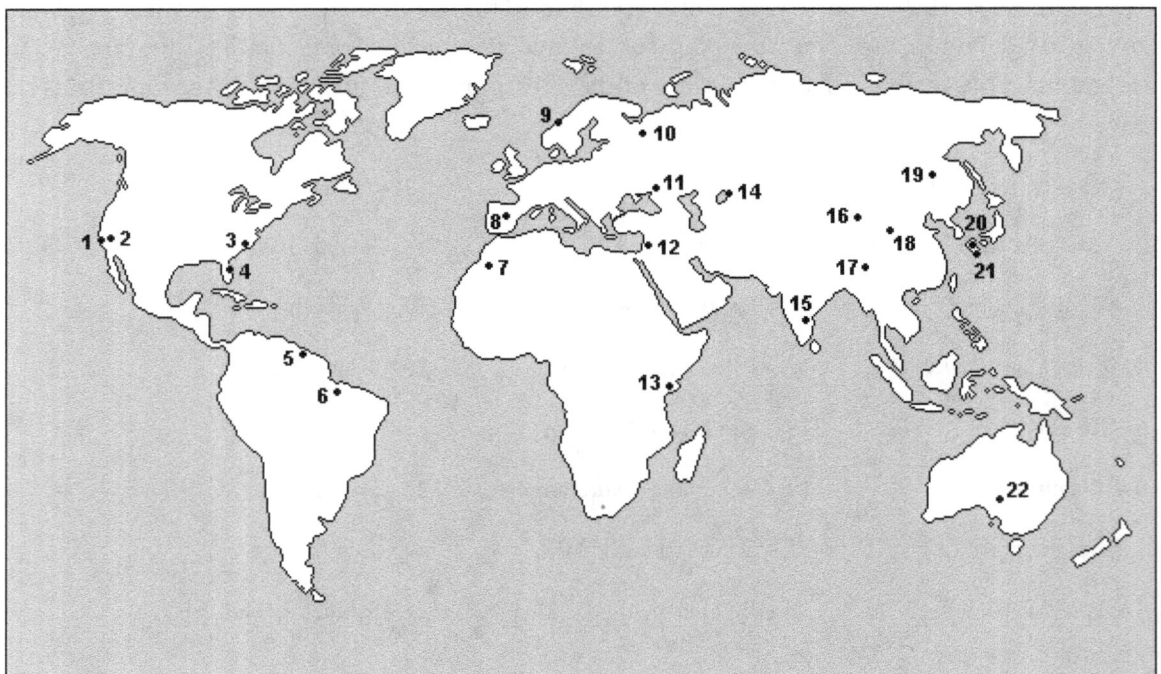

Figure 6-11: World Wide Launch Sites

Kennedy Space Center

Kennedy Space Center is located on Merritt Island (28 N, 80 W), just to the north of Cape Canaveral. It is NASA's primary launch base. It is operated by the National Aeronautics and Space Administration (NASA). Considerable support is provided by the Air Force Space Command's Eastern Space and Missile Center.

Western Range

The Western Range is located at Vandenberg AFB, California (35° N, 121° W). It is responsible for the missile and space launches from Vandenberg AFB and the Western Test Range which extends westward over the Pacific Ocean and into the Indian Ocean where it meets the Eastern Test Range. The nearest landmass directly to the south of Vandenberg is Antarctica. For this reason, launches to the south into polar orbits can be safely made. Surveillance satellites, low earth orbit (LEO) weather satellites, and environmental and terrain monitoring satellites like Landsat are launched from this facility. The safety limit of the launch azimuth is from 158° to 201° for an initial orbit inclination of 70° to 104°, respectively. There are approximately 52 launch pads, silos, and other sites to support launching the entire family of military and commercial rockets and missiles. The principal launch vehicles supported are the Delta, Atlas, and Minotaur.

Wallops Flight Facility

NASA operates the Wallops Flight Facility (38° N, 76° W), located on Wallops Island on the Atlantic coast, a few miles south of the Maryland and Virginia border. The principal activity now is the launch of sounding rockets although 21 satellites have been launched using the Scout launcher. Italian missile crews who launched Scouts from their facility off the coast of Kenya were trained here. Privately funded launches for commercial lifters (such as Scout and Conestoga) have been negotiated.

Poker Flats, Alaska

Poker Flats Research Range (65° N, 147° W), located northeast of Fairbanks, is owned and operated by the Geophysical Institute, University of Alaska, Fairbanks. It has the distinction of being the world's only university-owned launch range. Established in 1968, the range launches between ten and fifteen major sounding rockets, and a number of meteorological rockets, annually. Total launches to date are approaching three hundred. It also supports continuous ozone measurements and observations. NASA provides various range support radar and tracking systems and facilities.

Kodiak Island Alaska Spaceport

Alaska Spaceport (57.5° N, 153° W) is a commercial launch facility on 3,100 acres of Kodiak Island, Alaska, that can launch small satellites into polar orbit. It also provides a backup launch facility for Vandenberg Air Force Base for satellites needing delivery to polar orbit.

White Sands Missile Range, N M

Located at 32° N, 106° W. White Sands Missile Range was established July 9, 1945, as White Sands Proving Ground. It was the site of the July 16, 1945, Trinity shot, the world's first test of atomic bomb, and of postwar test and experimental flights with captured German V-2 rockets. It was also the scene of the February 24, 1949 launch of a Bumper rocket, whose second stage achieved altitude of 244 miles--

becoming the first man-made object in space. White Sands is now used for launches of suborbital sounding rockets and some anti-ballistic missile testing. New Mexico is in the process of establishing a spaceport adjacent to White Sands for conducting commercial orbital launches.

Commercial Launch

Sea Launch

The Sea Launch venture was formed in April 1995 in response to a growing market for a more affordable, reliable, capable and convenient commercial satellite launch service. A partnership was formed between Boeing Commercial Space Company, RSA Energia of Russia, NPO Yuzhnove of the Ukraine and Kvaerner of Norway to build and market a sea launch facility. Long Beach, California was selected as the home port and ground breaking for facilities began in August 1996.

Two unique ships form the marine part of Sea Launch. The Assembly and Command Ship is an all-new, specially designed vessel that serves as a floating rocket assembly factory as well as provide crew and customer accommodations and mission-control facilities at sea. The Launch Platform is a modified ocean oil drilling platform and is a self-propelled, semi-submersible launch complex which houses the integrated launch vehicle in an environmentally controlled hanger during transit to the launch site.

Chosen to capitalize on the Earth's rotation, the launch location maximizes Sea Launch's performance. The primary launch site will be along the Equator in international waters of the Pacific Ocean about 1,000 miles south of Hawaii. The selected booster for Sea Launch is the Ukrainian/Russian Zenit. Modified to a three-stage configuration, the Zenit was selected due to its highly automated launch procedures. In October 1999, Sea Launch made its first commercial launch using a Zenit-3SL.

Space Exploration Technologies Corporation

SpaceX, is an American space transport company that operates out of Hawthorne, California. It was founded in 2002 by former PayPal entrepreneur Elon Musk. It has developed the Falcon 1 and Falcon 9 space boosters, both of which are built with a goal of becoming reusable launch vehicles. SpaceX is also developing the Dragon spacecraft to be flown into orbit by Falcon 9 launch vehicles.

Launch Systems

There are only a few countries that have launch vehicles. Most launch vehicles are now available for commercial and military satellite launches.

Athena

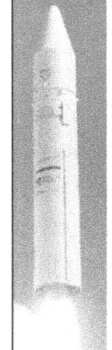

The Athena program was begun in January 1993. The first operational mission of the Athena, an Athena I, successfully launched the NASA Lewis satellite into orbit from Vandenberg Air Force Base (VAFB), CA., on Aug. 22, 1997. The first Athena II was successfully launched from Cape Canaveral Air Station (CCAS), Fla., on Jan. 6, 1998, sending NASA's Lunar Prospector spacecraft on its mission to study the moon. An Athena I also successfully launched the Republic of China's ROCSAT-1 satellite from CCAS on Jan. 26, 1999.

Figure 6-1:

The Athena I is a two-stage launch vehicle with the performance to place up to 1,750 lbs into a low earth orbit. The Athena II, a three-stage vehicle, more than doubles the payload capacity of the Athena I and has the performance to place up to 4,350 lbs into low earth orbit.

Delta II

The Delta II launch vehicle evolved from the Thor intermediate range ballistic missile (IRBM). It was first launched to carry satellites into orbit in May 1960. The Delta family of launch vehicles has gone through many upgrades and is available in a variety of configurations, depending on the needs of the customer. There are launch facilities at Vandenberg AFB and Cape Canaveral. The current configuration has a payload capacity of about 11,100 lbs (5,045 kg) into LEO or up to 4, 120 lbs (1,800 kg) into geostationary transfer orbit. The Delta II is the primary launch vehicle for the Navstar GPS satellites, a variety of US DoD, civil and foreign communications satellites such as Globalstar and some scientific payloads. In 1993, the Air Force designated the Delta II as its Medium Launch Vehicle (MLV-3).

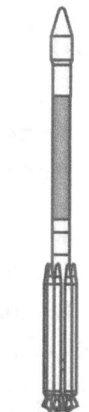

Evolved Expendable Launch Vehicle (EELV)

Figure 6-3: Delta

Evolved Expendable Launch Vehicle, known as EELV, is designed to improve the United States' access to space by making space launch vehicles more affordable and reliable. The program is replacing the existing fleet of launch systems with two families of launch vehicles, each using common components and common infrastructure. The vehicles are the Boeing Delta IV and Lockheed Martin Atlas V. EELV's operability improvements over current systems include a standard payload interface, standardized launch pads and increased off-pad processing.

As the Air Force's space-lift modernization program, EELV was designed to reduce launch costs by at least 25 percent over heritage Atlas, Delta and Titan space launch systems.

The Delta IV family of launch vehicles is designed for optimum performance for a wide range of flight profiles, and is capable of carrying payloads ranging from 4,231 kg (9,327 lb) to 12,757 kg (28,124 lb) to geosynchronous transfer orbit (GTO). The Delta IV Medium, Medium-Plus and Heavy configurations are evolved, combining highly reliable, flight proven systems from Delta II and III, while incorporating the latest technology into a family of vehicles maximizing the use of common hardware.

Commonality between all of the systems is central to the Delta IV. Each Medium & Medium-Plus vehicle uses a single common booster core (CBC), while the Heavy uses three CBCs. The Boeing Rocketdyne-built RS-68, a liquid hydrogen/liquid oxygen engine that produces 663,000 lbs of liftoff thrust, powers the first stage. This engine is mounted to the CBC first-stage structure and was designed for ease of manufacture by significantly reducing part count and thereby increasing reliability. Thirty percent more efficient than conventional liquid oxygen/kerosene engines, the RS-68 is environmentally friendly, producing steam as a combustion by-product. The three Delta IV Medium-Plus vehicles use a single CBC and are augmented by either two or four 1.5-meter (60-inch) diameter solid rocket strap-on graphite epoxy motors (GEMs).

The cryogenic second stage is an evolutionary design incorporating the Redundant Inertial Flight Control Assembly (RIFCA) from Delta II and the Pratt & Whitney RL10B-2 engine. The Delta IV Medium & Medium-Plus (4,2) vehicles use the same 4-meter diameter second stage, while the Delta IV Medium-Plus (5,2), Medium-Plus (5,4) and Heavy vehicles use the same RL10B-2 engine, but have larger 5-meter diameter fuel tanks and stretched oxidizer tanks.

On the Delta IV Medium & Medium-Plus (4,2), the payload is encapsulated in a 4-meter (13.1-feet) diameter payload fairing (PLF) for protection. On the Delta IV Medium-Plus (5,2), Medium-Plus (5,4) and Heavy, the payload is encapsulated with a similar 5-meter (16.7-feet) diameter payload fairing. Both the 4 and 5-meter diameter PLFs are composite bisector structures that were evolved from the Delta II 2.9 meter diameter and the Delta III 4-meter diameter PLFs. The Heavy vehicle can also employ a 5-meter diameter aluminum trisector fairing with Titan IV heritage.

Boeing has successfully launched seven Delta IV launch vehicles. Delta IV's inaugural flight was marked by the successful launch of a commercial satellite on a Medium-Plus (4,2) in November 2002. Two Air Force communication satellites were successfully launched on Delta IV Medium vehicles in 2003, and the first Heavy vehicle was launched in December 2004

The Lockheed Martin Atlas V resulted from Lockheed Martin's combination of the best practices from both the Atlas and Titan programs into an evolved commercial and government launch system for the 21st century. Atlas V builds on the design innovations demonstrated on Atlas III and incorporates a structurally stable booster propellant tank, enhanced payload fairing options and optional strap-on solid rocket boosters

All nine Atlas V to date were successfully launched. The Atlas V family uses a single-stage Atlas main engine, the Russian RD-180 and the newly developed Common Core Booster (CCB)TM with up to five strap-on solid rocket boosters. The CCBTM is 12.5 ft. (3.8 m) in diameter by 106.6 ft. (32.5m) long and uses 627,105 lbs. (284,453 kg) of liquid oxygen and RP-1 rocket fuel propellants

Additionally, on Atlas V, Lockheed Martin introduced a 4.57-meter usable diameter Contraves payload fairing in addition to retaining the option to use the heritage Atlas payload fairings. The Contraves fairing is a composite design and is based on flight proven hardware. Three configurations will be manufactured to support Atlas V. The short and medium length configurations will be used on the Atlas V 500 series

The Centaur upper stage uses a pressure stabilized propellant tank design and cryogenic propellants. The Centaur stage for Atlas V is stretched 5.5 ft (1.68 m) and is powered by either one or two Pratt & Whitney RL10A-4-2 engines, each engine developing a thrust of 22,300 lbs. (99.2 kN). Operational and reliability upgrades are enabled with the RL10A-4-2 engine configuration. The inertial navigation unit (INU) located on the Centaur provides guidance and navigation for both Atlas and Centaur, and controls both Atlas and Centaur tank pressures and propellant use. The Centaur engines are capable of multiple in-space starts, making possible insertion into low-earth parking orbit, followed by a coast period and then insertion into GTO.

Pegasus

Pegasus is a winged, three-stage rocket, launched from under the wing of an aircraft. It weighs about 41,000 lbs. and is 50.9 feet long with a 22-foot wingspan. The first launch was in 1990. The Pegasus launch vehicle is carried aloft by the L-1011 "Stargazer" aircraft to a point approximately 40,000 feet over open ocean areas, where it is released and then free-falls in a horizontal position for five seconds before igniting its first stage rocket motor. With the aerodynamic lift generated by its delta wing, the small rocket achieves orbit hundreds of miles above the Earth in approximately ten minutes. Most recently it has been used to launch the ORBCOM communication satellites. Advantages of using the Pegasus include:

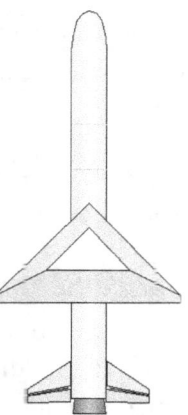

- Flexibility - Capable of being launched from virtually anywhere in the world with appropriate range facilities

- Cost - Approximately half the cost of equivalent ground-based launchers

Figure 6-4:
Pegasus

- Performance - Equatorial orbit up to 1,100 lbs., polar and sun synchronous orbits up to 750 lbs., geosynchronous transfer orbit up to 400 lbs., and Earth escape up to 300 lbs.

Applications- Supports a wide range of missions, including space technology validation, Earth science and space physics experiments, hypersonic flight research, Earth imaging, communications, and planetary exploration.

Minotaur

The Minotaur is combination of the first and second stages of the Minuteman II. The third and fourth stages are from the Pegasus XL launch vehicles. The rocket is part of an Air Force effort to use surplus Minuteman II components for sub-orbital and orbital spacelift. As of January 2004 the Air Force has about 350 Minuteman II ICBMs in storage. The vehicle is capable of launching several payloads of up to 750 lbs to a 400 nm, sun-synchronous orbit which is roughly 1.5 times the Pegasus XL capability. The first launch was in January 2000 when it was used to launch several microsatellites and picosatellites into orbit. The Minotaur IV uses decommissioned Peacekeeper ICBMs as the solid booster to accommodate a much higher payload capacity,

Taurus

Taurus is a four-stage vehicle. It is capable of launching a 2,100-lbs. payload into polar orbit. Taurus fills the cost and performance gap between our Pegasus rocket and the industry's much larger, more expensive launch vehicles. In March 1994, the Taurus rocket made its debut, placing two satellites for the Defense Advanced Research Projects Agency (DARPA) into near-perfect orbits. It is compatible with the USAF's Western Range (WR) and Eastern Range (ER) and NASA's Wallops Island Range 30-day on-pad hold capability. The Taurus was most recently launched in May 2004.

SPACE SYSTEM SEGMENTS

Space systems have three distinct segments:

- Space Segment: The satellites placed into orbit or components used to launch the satellites.

- Control Segment: The personnel, equipment and facilities responsible for the operation and control of the satellite and, in many communications systems, control of users' transmissions through the satellites.

- User Segment: The personnel, equipment and facilities that use the capabilities provided by the satellite payload.

Space Segment

There are numerous types of space systems providing a wide variety of capabilities and services. In spite of this diversity, there are similarities among all satellites because they all must operate in the environment of space. All satellites have two principal subsystems:

- The platform

- The payload

Platform

The platform is the basic frame of the satellite and the components which allow it to function in space, regardless of the satellite's mission. The control segment on the ground monitors and controls these components. The platform consists of the following components:

Structure of the Satellite

The structure or body of a satellite holds all of the components together as an integral unit and provides the interface with the launch vehicle. The structure must be strong enough to withstand the rigors of launch yet light enough to not unduly restrict payload weight. Many different shapes and materials have been used. Most satellites are built in low quantities. Each is designed to accomplish specific functions using technology and materials available at the time. As technology develops, new or improved materials and components are used to build new replacement satellites. It is not uncommon, therefore, that follow-on systems are significantly different in design and configuration even though they perform functions identical to earlier systems.

Power

Satellites require power to operate the electrical equipment that is on board. Satellites which have high power sensors or transmit strong or continuous signals require more power than those which have low power sensors and radios. For example, a satellite with a radar emitter and receiver requires a significant amount of power. Communications satellites which receive, process, amplify, and retransmit signals sent from users on the Earth or from other satellites require more power than a scientific satellite with only few sensors and a small radio to transmit the data to researchers on the ground. There are three types of energy sources:

- Solar Energy

- Chemical Energy

- Nuclear Energy (note we self impose restriction of nuclear propulsion for satellites in earth orbit)

Propulsion

Most satellites have an on board propulsion system which is used to achieve initial orbit and to make position changes. Shortly after reaching the initial, or transfer, orbit the satellite is separated from the final stage of the launcher. The final orbit is achieved by firing a kick motor to move the satellite into the final desired orbit and position. Some satellites are designed so that they can be repositioned. In general, changing the orbital plane requires more force than changes within the orbital plane. The kick motor is used to make major changes in the satellite's orbit. Sufficient propellant must be carried on board to last the lifetime of the satellite system. After a satellite's payload is no longer useable, the kick motor is often used one last time to either increase or decrease its orbital velocity. Increasing the speed raises the altitude of the orbit. Decreasing the speed lowers the altitude, sometimes enough so that the satellite is deliberately destroyed reentering the Earth's atmosphere. A replacement satellite will then be able to assume the vacated orbital position.

Stabilization and Attitude Control

Stabilization and attitude control are necessary to ensure that the satellite maintains the proper attitude. Satellites are subjected to a number of forces in space such as particles streaming from the Sun, meteorites, atmospheric drag, gravity from the Moon, gravity gradients and other perturbations. These forces cause satellites to wobble, spin, drift, or move in other ways not desired. Most satellites which provide visual or electronic images of the Earth or its environment maintain three axis stabilization (roll, pitch, and yaw). Many communications satellites are designed to rotate about their longitudinal axis (roll) and thus have only two axis stabilization. Two and three axis stabilization allow sensors and antennas to be pointed in specific directions. Devices such as momentum wheels on the satellite help to stabilize the satellite while in orbit. Position, velocity and attitude data from on-board sun sensors, star trackers, horizon scanners and other devices is transmitted to ground control stations. When momentum wheels and other such passive devices cannot compensate or adjust the orbit, the satellite controllers send signals to the satellite to fire thrusters in short spurts to control roll, pitch, yaw and to make corrects in orbital altitude. To reduce size, mass, complexity and cost some small satellites are designed to tumble freely through space without any stabilization or attitude control.

Thermal Control

The temperature in a satellite must be controlled so that components do not become too hot or too cold. The temperature in a satellite is affected by both internal and external sources. On board electronic equipment and other devices which consume power generate heat. The Sun is a source of a vast amount of radiant energy. Radiant energy absorbed by the satellite heats the satellite surfaces and components unless it is dissipated. The ambient temperature of space is a few degrees above absolute zero (-459 degrees F) however, since there is almost no atmosphere, heat transfer between the satellite and the space around it by convection or conduction is almost nonexistent. In very low Earth orbit, however, there can be significant heat generated from friction as the satellite moves at very high speed through the outer reaches of the atmosphere. The most common heat transfer device is the passive radiator which radiates heat from the satellite into space to maintain temperatures within design parameters. A satellite may also require a liquid or gas filled cooling system to transfer heat from internal components to the passive radiator.

Environmental Control

Manned spacecraft require precise environmental control to ensure that air quality, humidity, water and temperature are maintained within operating limits. Most unmanned satellites do not require environmental control.

Telemetry, Tracking and Command (TT&C)

The TT&C subsystem monitors and controls all of the other systems on the spacecraft, transmits the status of those systems to the control segment on the ground, and receives and processes instructions from the control segment. Telemetry components include sensors throughout the satellite to determine the status of various components, the transmitters and antennas to provide the data to the control segment and even the data itself. (In some documents TT&C stands for Telemetry, Tracking and Control. The tasks are, however, the same).

Payload

The function and capabilities of the payload are the reasons a satellite is placed in orbit. The payload provides space-based capabilities to the users. The payload distinguishes one type of satellite from another. The general types of satellite systems are:

- Communications

- Position/Navigation

- Reconnaissance, Surveillance and Target Acquisition (RSTA)

- Early Warning

- Weather and Environmental Monitoring

- Scientific/Experimental

- Manned

Control Segment

The control segment is responsible for the operation of the overall system which includes platform control, payload control and network control. The control segment consists of personnel, ground satellite control facilities and systems on the satellites.

Platform Control

Platform control involves satellite station keeping, relocation maneuvers and the proper functioning of onboard systems. Platform control must be accomplished for almost all satellites. The tasks involved are accomplished through TT&C.

Payload Control

Payload control involves operation and control of the payload on the satellite. Data is provided by the satellite and commands are sent to the satellite through TT&C.

Network Control

In general, there are two types of networks involved with a satellite system, both of which must be controlled:

- The network of satellite control and monitoring stations. Control and monitor stations are strategically located around the world to perform platform and payload control.

- The network of user terminals. Satellite systems which receive and process transmissions from multiple users (such as WGS, UFO and Milstar) require a controlled earth terminal segment (users) to insure that the system provides service to authorized users according to established priorities and within the operating parameters of the system. These systems can provide service to a limited number of users at any one time. Satellite systems which only transmit signals to users (such as GPS, DMSP and GOES) may have an unlimited number of users and do not require control of user terminals. More information on network control of user terminals is contained in the sections of this chapter pertaining to specific systems.

Telemetry, Tracking and Commanding (TT&C)

Satellites are controlled through TT&C. Some TT&C is required for all satellites, regardless of the payload they carry.

Telemetry

After the satellite has been placed in the desired orbit, there must be a way to make the data, which the sensors detect, available to the engineers or operator users in order for it to be meaningful. Such a method does exist and is referred to as telemetry. Telemetry is information transmitted to a ground station over a radio link that evaluates the satellite performance or provides mission data.

General spacecraft health and status data is fairly consistent, regardless of the type of mission. This data consists of pressure, temperatures, flow rate, voltages, current and events present throughout the satellite system, subsystems and components. Housekeeping instruments indicating switch positions are digitized, designating the "ON" or "OFF" modes as "1" and "0," respectively. These switches reflect satellite health and status information. Note that, when communicating with spacecraft, space holding bits ("S" bits) are sent whenever "1s" or "0s" are not being sent.

Tracking

Before we can communicate with a ballistic missile or orbiting satellite, we must know where it is with respect to our ground stations. Tracking is the process of making observations of the spacecraft's position relative to a tracking station or other fixed point whose position is accurately known. Orbit determination is the process whereby the tracking observations are used to determine the spacecraft's orbital characteristics and its position in space.

Commanding

Controlling a satellite and its functions from the ground is a crucial task, and is performed for most satellites. This task is accomplished by transmitting specially coded instructions from the ground station over radio frequency carrier, referred to as the uplink, to the satellite's receiving equipment. This entire process is known as satellite commanding and, depending on the manner in which the command is structured, will determine what functions will be performed.

Commands may be sent for accomplishing any of the following functions; ascent control, orbit adjust, reentry by separation, engine ignition or cutoff, control of internal systems, on-off, switchover, control of sequential events that must operate in a predetermined manner, or control of a space-borne timer which in turn controls a predetermined sequence of events. Commands falling into this latter category control such things as start time, orbital position for start, rate or event occurrence and on-off and reset of the timer.

Commands can be identified as either Real-time Commands (RTC) or Stored Programs Commands (SPC). The primary difference between these commands is the time of execution. RTC causes events within the satellite to occur upon command receipt while the satellite is still within sight of a ground station. The SPC is sent to the satellite while it is still within a ground station view, but it causes certain functions to be performed after the satellite has passed out of sight of a ground station.

It is significant to note, that on spacecraft which contain no self-sustaining reference package, the capability to be commanded is an absolute necessity. Vehicles in this category undergo continuous telemetry monitoring and rely solely upon commands sent from the ground, the bulk of them being of a station-keeping nature.

Air Force Satellite Control Network

The Air Force Satellite Control Network (AFSCN) is designed to have the flexibility to support a wide spectrum of orbiting satellites. A large number of satellites with various altitudes and orbit inclinations are supported on a 24-hour per day, 7-day a week schedule. In addition to the primary support provided to the Department of Defense (DoD), the AFSCN also provides services for non-DoD organizations, including NASA, and US sponsored programs of foreign governments.

The AFSCN consists of Mission Control Complexes (MCCs), Remote Tracking Stations (RTS), Automated Remote Tracking Stations (ARTSs), and test facilities located around the world. These personnel, equipment, and facilities of the AFSCN are used to accomplish satellite maintenance. However, these resources are not responsible for the payload data received from the various programs, and do not directly interface with the program users. The resources are there to maintain the satellite in the optimum orbit for mission accomplishment and to ensure the vehicle and payload are performing as designed. They also are there to assist in the restoration of malfunctioning satellites.

The AFSCN accomplishes its assigned mission by tracking data generation, telemetry monitoring, commanding, and testing.

- Tracking Data Generation -- Tracking Data is the range, range rate, azimuth, and elevation angles with a time tag received from the RTS. This data is used to determine the position of the satellite in its orbit.

- Telemetry Monitoring -- The AFSCN allows crews to monitor the telemetry data from satellites. From this data, the health of the vehicle's various subsystems/sensors can be determined.

- Commanding -- The AFSCN also allows crews to send commands to the satellite to operate/configure it. These commands can either be executed in real-time or can be stored by the vehicle for execution at a specified time.

- Testing -- Some communications and navigation satellites require extensive measurement, testing, and calibration while in orbit. This procedure is normally performed soon after launch during early orbit testing; however, it can also be done throughout the life or even at the end of life for data collection/trouble shooting purposes. The information gathered in this method can be used to improve future spacecraft.

The network consists of eight subordinate tracking stations located around the world: 23rd Space Operations Squadron, New Boston Air Force Station, NH; Detachment 1, Vandenberg AFB, CA; Detachment 2, Diego Garcia, Chagos Archipelago; Detachment 3, Thule AB, Greenland; Detachment 4, Kaena Point, Oahu, HI; Detachment 5, Andersen AFB, Guam; Colorado Tracking Station, Schriever AFB, and Oakhanger, England, operated by the United Kingdom. The tracking stations command, track, record, and process on-orbit satellite data in support of DOD, NASA, and NATO programs.

FORCE ENHANCEMENT

Space force enhancement functions are similar to combat support operations in that they improve the effectiveness of forces across the full spectrum of operations by providing operational assistance to combat elements. Command and support elements also integrate space force enhancement functions into their operations. As outlined in Joint Publication 3-14, the functions include:

- Satellite Communications

- Space-based Positioning, Navigation, and Timing

- Environmental Monitoring

- Missile warning

- Intelligence, Surveillance and Reconnaissance

Civil, commercial, and allied capabilities may augment DoD systems to support military space force enhancement requirements. The efficiencies resulting from the use of these space capabilities can have a dramatic effect on Army operations.

COMMERCIAL/CIVIL SPACE

Space-based systems and products will increase our adversaries' potency and level the military battlefield. The explosive commercial development of space capabilities will make space products accessible to any organization with resources. Amplifying the potential threat is the compression of time from "intent to use" because future space systems will operate in near real time.

As described in several instances above and throughout this text, our adversary has access to many space products and capabilities. The proliferation of space products throughout the commercial environment gives the enemy access to communications, high-resolution imagery, current weather, and accurate position/navigation/timing. National policy attempts to resolve some of the problems associated with the US commercial space products but does not go far enough. The commander of a military operation must be aware of what is available to his adversary, how it might be used, and what can be done to protect his own use of commercial assets while denying it to the enemy.

Communications

Technology advances provide unparalleled capability and access to potential adversaries in the form of higher bandwidths and a proliferation of global communications. An adversary's ability to command and control forces will gain much from dynamic, global, communication networks such as Iridium and Globalstar.

There are mechanisms that are being considered to give industry incentives to increase protection levels and decrease an adversary's access.

Imagery

There is concern about the availability of what formerly was highly classified imagery with very high resolution. But already on the market is one-meter and lower resolution multispectral and optical imagery. It is not available typically in near-real-time because it can take hours to image, receive and process the product. Had near-real-time capabilities been available in 1991 and if Iraq had had access to the capabilities, the left hook that the Coalition set up in DESERT STORM might not have worked because Iraq would have had imagery of the troop formation.

Anyone with Internet access will get highly accurate imagery almost instantly. A potential enemy can order archived imagery from any number of commercial agencies and can even request specific imagery be taken over any area of the Earth. Some of the websites available to him are:

- Space Imaging: http://www.spaceimaging.com/

- Canada Centre for Remote Sensing: http://ccrs.nrcan.gc.ca/index_e.php

- Google Earth: http://earth.google.com/

- Landsat Program: http://geo.arc.nasa.gov/sge/landsat/landsat.html

- Radarsat: http://radarsat.space.gc.ca

- EROS: http://edc.usgs.gov/

- Other: http://terraserver.com

There is commercial satellite tracking software available that anyone can purchase to track the path of any satellite have the ephemeris data for. Most data is available on-line also. Some of the software packages available are:

- Satellite Tracker: http://www.n2yo.com

- Heavens Above: http://heavens-above.com

- Northern Lights Software Associates: http://nlsa.com

- Satellite Toolkit: http://www.stk.com

Weather

All geostationary weather imagery is available to anyone via WEFAX or the Internet. Most of the polar orbiting satellites also produce weather imagery that is available to anyone. Depending on equipment the enemy has in theater, he may very well be receiving and analyzing the same weather picture that we are. DoD has passed management of national weather satellites to the National Oceanic and Atmospheric Administration except for the Defense Meteorological Satellite Program which will be discussed later. Cable and satellite television stations now routinely market satellite weather information.

Position/Navigation/Timing

Current US policy states we will give the commercial and civil sector more accurate position and navigation data and that GPS will be a global satellite system. Russian has renewed its efforts to update and profide a full GLONASS constellation and will be operational again in 2012. Galileo is the European community's space-based navigation system of the future. It is now scheduled to have 26 Galileo satellites in medium Earth orbit by late 2015, with the likelihood of six more to follow a year or two later to round out the in-orbit fleet to 30 operational spacecraft and two spares. The six final satellites will not be ordered until 2014, when the European Commission's new seven-year budget takes effect and provides Galileo with a fresh source of cash. Potential enemies will use all of these systems to enable their own precise military operations.

GPS jammers can be built for a relatively small amount of money. There are also jammers being advertised for sale with one 4-watt jammer being advertised states it has an effective jamming area of 150-200K. We have seen the use of GPS jammers during Operation Iraqi Freedom and Enduring Freedom. A capable adversary may be able to severely affect operations with this technology.

Chapters following will go into detail regarding the Space Force Enhancement areas.

Chapter 6
Satellite Communications

OVERVIEW

Communicating with military units and individuals dispersed across the battlefield or deployed a great distance from their home stations has always been a major challenge and our military continues to adapt to the political and military situations of the future. The requirement for immediately deployable forces has increased significantly. Forces require Beyond Line Of Sight (BLOS), long range communications while enroute, immediately upon arrival and continuously throughout and following deployment. Radio communications using Ultra High Frequency (UHF), Super High Frequency (SHF) and Extremely High Frequency (EHF) are limited to line-of-sight (LOS). High Frequency (HF) transceivers bounce signals off the ionosphere; however, the ionosphere is always changing, thus HF signals tend to fade in and out. An alternative is to connect into existing communications systems in the operational area. In many places, however, an adequate communications infrastructure does not already exist in the area of operations or it will take significant resources and time to connect into it.

Communications was the first practical application of space technology for both the military and business. Today there are more communications satellites in use than of any other type. Satellite communications provides the warfighting commander a reliable means to control his forces and conduct operations at various operational tempos. Satellite systems provide the backbone of national and DoD worldwide communications

Military Satellite Communications (MILSATCOM) provides single channel and multi-channel communications throughout the military. Multi-channel terminals provide range extension and joint service interoperability. Single channel terminals support battlefield voice and data communications as a part of Combat Net Radio and special communications for force management, Emergency Action Message (EAM) dissemination and Special Operations Forces (SOF) communications.

Satellite communications (SATCOM) systems can significantly enhance our joint communications capabilities, but they are not a panacea. Satellite communications do not necessarily replace but rather complement ground based communications systems. A thorough understanding of the operations, types, capabilities and limitations of military and commercial satellite communications systems is necessary for the military leader to use SATCOM to enhance operations.

COMMUNICATIONS

Orbits of Communications Satellites

Although the majority of communications satellites operate in geostationary orbits, there are other orbits that are used to provide coverage to other areas or to provide a capability not suitable for geostationary satellites. **Figure 6-1** shows three representative orbits typically used by communications satellites (annotated here as Type 1, 2 and 3). The resultant ground traces of each type are shown on the right.

GEO

Geosynchronous Earth orbits (GEO) (Type 1) are typically used to provide continuous communications to a specific area because the satellite's motion is synchronized with an area of the Earth below it along the equator. A geosynchronous orbit can have any inclination. A minimum of three satellites in geosynchronous orbit is needed to provide global coverage. There are some communications satellites in GEO, although most are in a special GEO orbit called geostationary.

As discussed in Chapter 4, a geostationary orbit is a type of geosynchronous orbit which seems to be

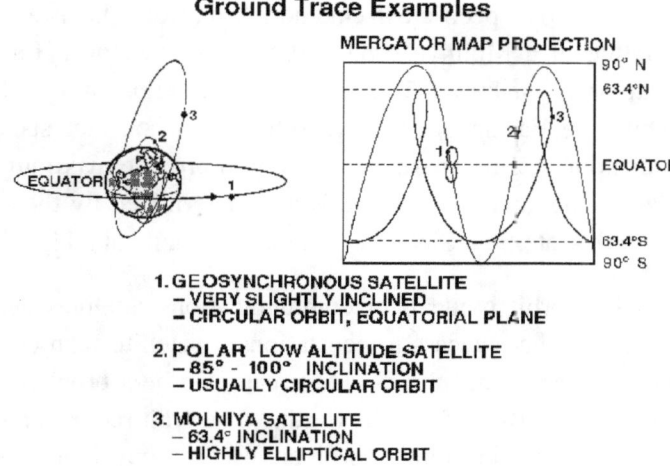

Ground Trace Examples

1. GEOSYNCHRONOUS SATELLITE
   – VERY SLIGHTLY INCLINED
   – CIRCULAR ORBIT, EQUATORIAL PLANE

2. POLAR LOW ALTITUDE SATELLITE
   – 85° - 100° INCLINATION
   – USUALLY CIRCULAR ORBIT

3. MOLNIYA SATELLITE
   – 63.4° INCLINATION
   – HIGHLY ELLIPTICAL ORBIT

Figure 6-1: Types of Orbits and Ground Traces

positioned at one location over the equator at an altitude of approximately 25,645 statute miles (22,300 nautical miles) at 0 degrees inclination. In practice, the coverage of geostationary satellites is limited to areas between approximately 70° north latitude and 70° south latitude. For our ground forces this has limited impact since only the very northernmost part of Alaska, northern Greenland, the northern tip of Norway, northern Siberia, and Antarctica are in the area not covered by geostationary satellites. However, it does impact naval forces and some strategic Air Force missions. Geostationary orbits have the communication advantage of a large, stable footprint. Fixed antennas can be used without tracking the satellite. This orbit is currently the world's standard for most GEO communications satellites. There are already many satellites positioned in the GEO belt. It will become increasingly difficult in the future to de-conflict GEO belt locations for new satellites because of overcrowding demand.

LEO

A communications satellite in low Earth orbit (LEO), polar, (Type 2) passes within view of every point on the Earth at some time during the day. A LEO in any other inclination will not cross the poles. LEO satellites offer many advantages for communications. LEOs are normally used for short burst, narrowband communications using radio frequencies below one gigahertz. Time delay is decreased for communications traffic using LEO satellites because the satellites orbit closer to the Earth which also allows smaller and simpler antennas to be used. These low altitude communication satellites require less power to transmit signals because of the shorter signal path.

The biggest disadvantage of LEO as a communications orbit is each satellite is not in view of the ground receiver at all times. Their footprint is small and quickly passes out of view of the ground terminal. Because of this, communication satellite constellations in LEO normally consist of several satellites to provide 24-hour coverage.

HEO

The Highly Elliptical Orbit, specifically the Molniya orbit (Type 3), is designed to provide extended coverage to a specific extreme northern area. This is a highly elliptical orbit (HEO) inclined to 63.4°. A satellite in a Molniya orbit, a USSR derived orbit for communications, is within view of its ground stations about 80% of the time. As a satellite in a Molniya orbit approaches its apogee it slows down while the Earth continues to rotate at a constant speed. The resultant ground trace makes a loop, indicating that this particular orbit will provide maximum coverage time over the northern latitudes. The size of the area of coverage is also determined by the width of the antenna beam and signal frequency from the satellite. Since the satellite is usually at a high altitude, the area of regard can be large.

This orbit is ideal for communications satellites used to provide coverage in the extreme northern latitudes where access to geostationary satellites can be difficult. Molniya orbits have poor coverage of the southern hemisphere. There can also be a problem in transmission delays and signal loss when the satellite is farthest from the Earth. The orbit requires multiple satellites for 24-hour coverage. Normally three satellites in Molniya orbit are used for coverage.

MILITARY COMMUNICATION APPLICATIONS

SATCOM resources are essential for military force projection. The seamless connectivity provided by SATCOM ensures that planning and coordination occurs when and where needed.

The Army uses single channel SATCOM to support tactical battlefield voice and data range-extension requirements for non-contiguous battlespace operations. The ground terminals are user owned and operated and can be man portable, man-packable, or vehicular mounted. Single channel SATCOM is used extensively in a variety of warfighter missions because of the flexibility and mobility provided. Military single channel SATCOM is used to provide communications for SOF operations, mobile warfighter C2 nets, and intelligence dissemination.

The primary purpose of multi-channel SATCOM is to extend the range of the area common user system (ACUS). Strategic ground terminals provide the international satellite trunking for the Defense Information Systems Network (DISN). Tactical terminals are deployed to provide range extenuation links for critical nodes and strategic reachback.

Global Broadcast Service (GBS) is a tool for the warfighter that provides a high speed, one-way flow of information tailored to a commander's mission requirements. GBS provides battlefield awareness video and data broadcasts down to brigade level (i.e. Unmanned Aerial System (UAS) video) plus any other theater intelligence broadcasts necessary to support operations.

The use of commercial SATCOM is evident throughout the Army. Commercial SATCOM is an alternative means of satisfying those communications requirements that cannot be satisfied using MILSATCOM. INMARSAT and INTELSAT are a couple of examples of commercial SATCOM systems that the Army has relied upon for communications. Mobile Satellite Service systems such as Iridium are rapidly assuming a place in the Army SATCOM architecture. Currently, commercial SATCOM (vice MILSATCOM) is used primarily for Admin and Log C2, as a surge supplement for

MILSATCOM, to provide Mobile Satellite Service with cellular-like voice and fax capability, and for peace keeping, disaster relief and humanitarian operations.

Each phase of an operation brings with it a unique communications plan. Early in the deployment scenario, there is a heavy reliance on single channel SATCOM. As the operation progresses and more units are established in country, more emphasis is placed on multi-channel SATCOM. Peacetime operations require a different mix of communications and SATCOM requirements than wartime operations. The need for C2 links that are reliable, fast and flexible dictate the use of SATCOM. SATCOM must become a seamless part of the overall communications network that will successfully function and interoperate in a complex and sensitive single-nation or multinational mission.

Satellite communications provided invaluable support during Operation DESERT STORM. About 75% of the more than 1,500 satellite communications terminals deployed to the theater were single channel, man portable military and commercial sets. They provided critical inter and intra-theater communications links to widely dispersed and fast moving units.

Task Force Hawk used SATCOM during operations in Kosovo to achieve proper C2 for the AH-64s. The Task Fore Commander determined early on that the mountainous, rugged terrain would require a robust C2 plan. SINCGARS and HAVEQUICK II were not effective in this environment because of limited line of sight. A UH-60 with SATCOM was used to relay messages to Hawk Base. The recommendation after the operation was the AH-64 should be equipped with SATCOM.

In operations in Somalia and Afghanistan, SATCOM was essential to deployed forces due to a combination of their relatively small size, wide disbursement, and the lack of a local communications infrastructure.

In Operation Iraqi Freedom, SATCOM is used to provide fast moving ground forces with communications. The organic Army ground communications systems were unable to deploy and set up fast enough to be able to support the fast moving advance.

Other changes in operational capability have been implemented through the use of SATCOM. For example, UAS are able to fly over enemy held areas and send video and sensor data over SATCOM. There are numerous examples of ground soldiers and Forward Air Controllers using SATCOM to communicate directly with in-bound aircraft to reprogram the target coordinates before the aircraft were within line-of-sight communications range. The effect was that long range bombers were able to be quickly redirected to hit targets of opportunity and provide close air support to ground forces because of the use of SATCOM.

SATCOM Considerations

Frequency Ranges

The frequency of any communications signal is the number of cycles per second at which the radio wave vibrates or cycles. Electro-magnetic waves cycle at phenomenal rates: one thousand cycles per second is called a kilohertz (kHz), one million cycles per second a megahertz (MHz), and one billion cycles per second a gigahertz (GHz). Today we refer to the continuum of frequencies used to propagate communications signals, 100 kHz to 100 GHz and beyond, as the electro-magnetic spectrum. SATCOM resides primarily in the UHF to EHF range.

Figure 6-2a shows the communications applications portion of the frequency spectrum. The top portion highlights various transmission mediums. The lower portion shows the wave propagation effects. Radio waves that propagate along a line of sight are best suited for satellite communications. Ground wave, surface wave and ionospheric reflection are not appropriate for satellite communications

Communication Applications

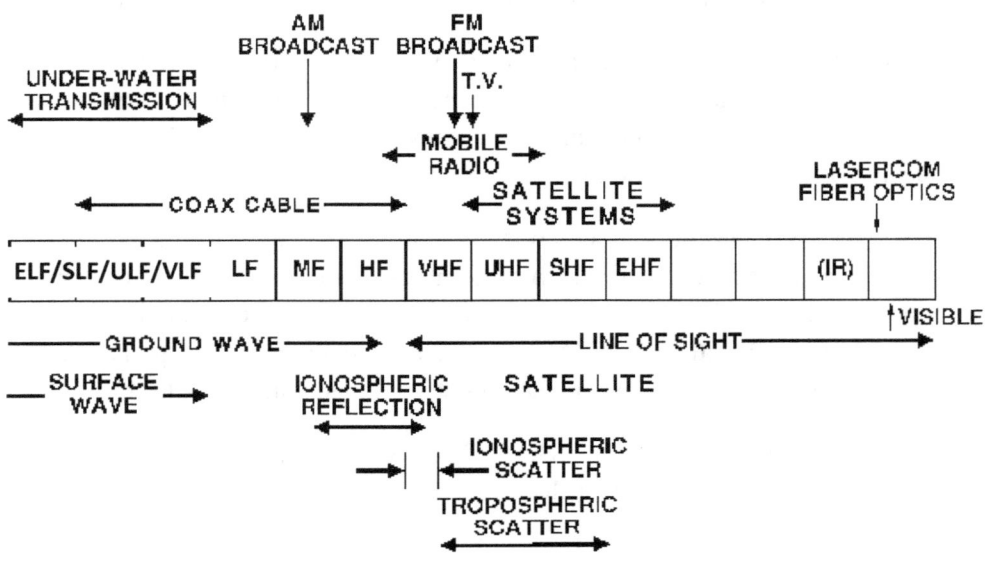

Figure 6-2a: Frequency Bands and Communication Applications

Typically, both military and civil SATCOM frequencies include Ultra High Frequency (UHF) (0.3 – 3 GHz), Super High Frequency (SHF) (3 – 30 GHz) and Extremely High Frequency (EHF) (30 – 300 GHz) bands of the frequency spectrum. **Figure 6-2b** points out that the designations of frequency bands can be confusing. The US military defines the frequencies of the UHF, SHF and EHF bands differently than the International Telecommunication Union (ITU). The Institute of Electrical and Electronics Engineers (IEEE) uses letters to designate frequency bands. The military UHF SATCOM band actually covers frequencies from 225 MHz to 400 MHz. Technically the frequencies below 300 MHz are in the VHF band according to ITU standards, but for military purposes we still refer to it as UHF. The military SHF SATCOM band is often referred to as the military X-band which covers 7.25 GHz to 8.4 GHz. The lower portion of the military X-band within the SHF band is actually in the upper portion of the C-band as defined by the ITU. The military EHF SATCOM band covers frequencies 22 GHz to 40 GHz. The lower portion of the military EHF band is in the upper portion of SHF band as defined by the ITU.

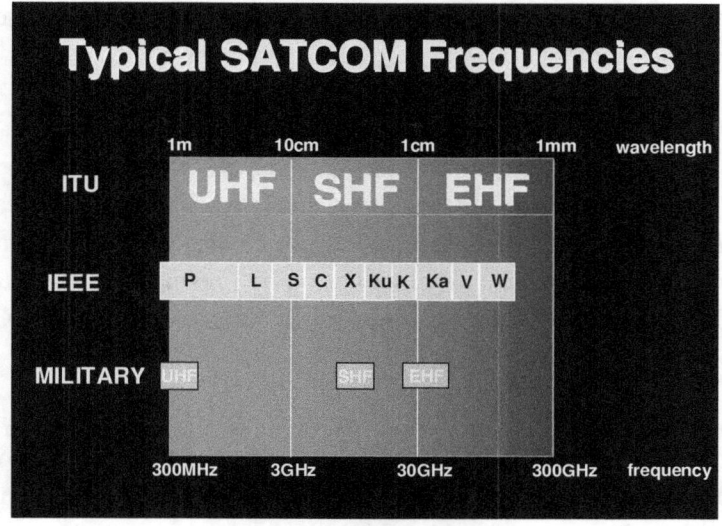

Figure 6-2b: Typical SATCOM Frequencies

Ultra High Frequency (UHF)

A UHF communications satellite uses the lowest frequency of all the MILSATCOM systems. Military UHF falls in the band of frequencies from 225 MHz to 400 MHz. It has a narrow bandwidth and a wide beam width. It provides only a relatively low data rate and single channel voice communications to small, portable transceivers. The wide beam width of the signals means that antenna pointing is not as critical as with higher frequency systems.

UHF is the only part of the spectrum the permits sufficient triple canopy foliage signal penetration to small hand-held terminals. Also, there is a great deal currently invested in UHF terminals and any follow-on system will have to take into consideration the need for backward compatibility.

It has only limited resistance to jamming because the frequency range and bandwidth is narrow; therefore, there is less that can be done to counter jamming or interference and still maintain communications. The impact of space weather on the ionosphere can degrade UHF communication signals. UHF is adequate for voice circuits or low data rate transmissions. UHF user terminals are relatively inexpensive, simple to operate and do not require large antennas. All the US military services use UHF, especially the Navy and the Army.

Super High Frequency (SHF)

SHF communication satellites are the most common type. They form the backbone of the space portion of the Defense Communications System. The higher frequencies result in a narrower beam width which allows the transmitted power to be concentrated into a smaller area. The military commonly uses the term "X-band" to mean the specific band of frequencies from 7.25 – 8.4 GHz that are strictly for military use. Future SHF SATCOM systems will use even higher frequencies in the Ku-band. The higher the frequency, the greater the bandwidth and therefore allows for more data transmission. Additionally, more anti-jamming techniques can be used. Sophisticated modulation techniques can be used to increase the number of simultaneous users. There is more atmospheric attenuation than at lower frequencies. The Army has a significant number of SHF satellite terminals.

Extremely High Frequency (EHF)

EHF communications satellites are the newest type. Typically the military uses the Ka and V-band. The high frequency allows a wide bandwidth and a very small beam width. The beam width can be small enough so that the transmitted signal is focused into a spot beam. Very high gain antennas are used to receive these EHF signals. Considerable anti-jamming can be implemented to assure communications in even the most demanding jamming conditions. Atmospheric attenuation, especially from clouds and rain, can be significant.

There is one major drawback to satellites down linking signals at frequencies greater than 10 gigahertz: the length of these microwaves is so short that rain, snow or even rain-filled clouds passing overhead can reduce the intensity of the incoming signals. At these higher frequencies, the length of the falling rain droplets are close to a resonant sub-multiple of the signal's wave length; the droplets therefore are able to absorb and de-polarize the microwaves passing through the Earth's atmosphere. In places such as Southeast Asia or the Caribbean, torrential downpours can lower the level of the incoming Ku-band satellite signal by 20 dB or more. This may severely degrade the quality of the signals or even interrupt reception entirely. Predictability of rain outages in certain parts of the world can at least be planned for as a consideration.

Bandwidth

Frequency, first and foremost, along with power and gain, dictate actual usable bandwidth capacity. Satellite design and use is directly tied to how much channel capacity, or information throughput, a satellite transponder can accommodate. Throughout the satellite communications industry the term "wide bandwidth" is commonly used to mean "high throughput capacity" and "high channel capacity", although (strictly speaking) these terms are not necessarily equivalent. The greater a transponder's bandwidth, the greater will be its potential channel capacity (in bits per second) to convey information at higher throughput rates. However, wider bandwidth (like found in EHF) means an increase in noise. Unless the conditions in all parts of the transmission signal path can be made just right to minimize the noise or to make the signal more powerful than the noise, the wider bandwidth will not allow for increased capacity. Transmitted signal power, the gain of the antennas and the efficiency of the receiver all have an impact on throughput capacity.

Antenna Size

Ground antennas come in different sizes and shapes and are designed for a specific purpose. If the dish is large then it will pick up more signals from space and be able to transmit with less power. If the dish is small it will pick up fewer signals from the satellite and require more power. The size of an antenna is determined by the frequency (wavelength) of the transmitted signal, the bandwidth (rate of transmission) and the desired gain or signal strength. Other factors remaining constant, larger antennas are needed at lower frequencies which have longer wavelengths. Higher bandwidth at a particular frequency requires a larger antenna. Digital messages transmitted at a slow rate do not require high gain to provide reliable communications. Voice circuits generally don't need high gain. High speed data circuits must have superior reliability so that the transmitted data is not altered.

Susceptibility to Jamming

Any receiver can be jammed. Satellite systems are no exception. In two way satellite communications systems there are receivers and transmitters on the ground and on the satellites. Susceptibility to jamming is determined by the frequency, bandwidth, signal modulation and other factors. Lower frequencies have narrower bandwidth and are more susceptible to jamming. Beam width of an antenna is inversely proportional to the frequency, thus lower frequencies have a wider beam width. Higher frequency transmitters can limit the beam width of the transmitted signal to a more specific area. More effective signal modulation techniques can be used at higher frequencies.

Most user receivers use directional antennas to communicate with the satellites. This limits the strength of jamming signals being received unless the jammer is in the path of the signal from the satellite to the receiver. Since the antennas of ground receivers are pointed into the sky, they are not significantly susceptible to ground jammers unless they are extremely powerful or very close.

Commercial satellites have virtually no protection against jamming. Military communications satellites have some ability to resist jamming through the use of various signal modulation and some frequency hopping techniques. Although jamming can be countered, a common result is a decrease in the amount of traffic that can be transmitted through the system when anti-jamming techniques are implemented. It is a conscious decision by the military to exchange data rate for security.

Solar activity such as large solar flares can disrupt all long distance radio transmissions.

Capabilities by Frequency

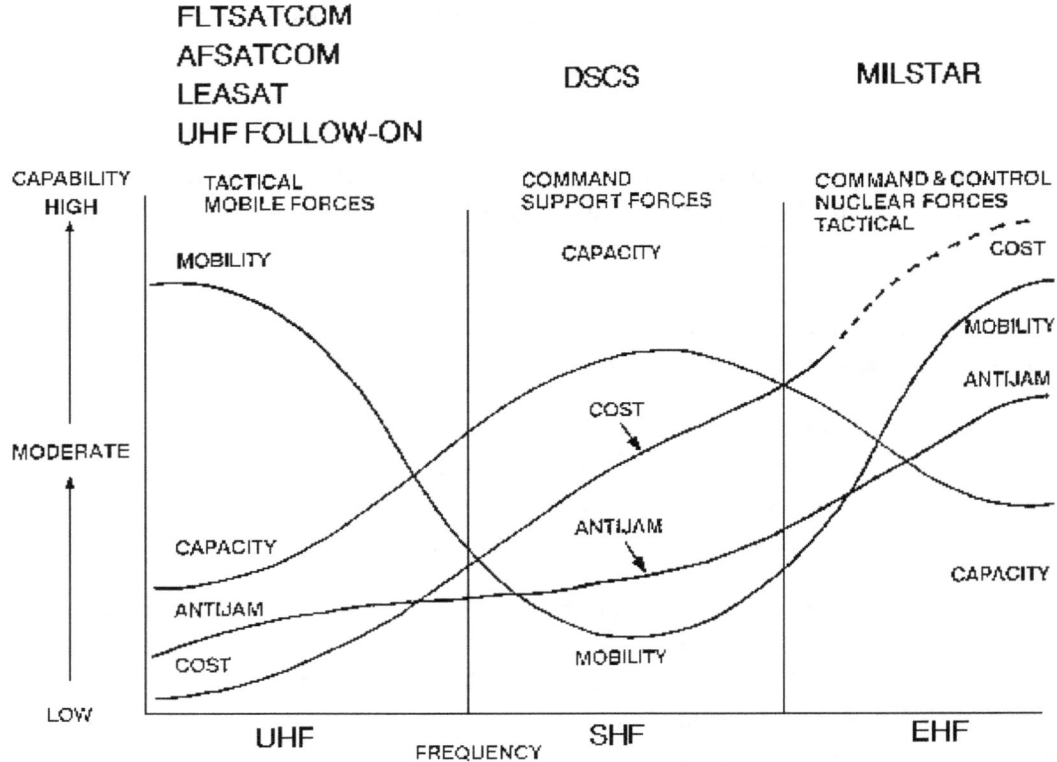

Figure 6-3: Capabilities and Limitations of UHF, SHF, EHF

As indicated in **Figure 6-3**, SATCOM using UHF frequencies are normally more flexible and mobile than the other frequency ranges. SATCOM in the UHF frequency can operate in dense overhead cover and rain. It is the least costly and therefore used quite frequently. SHF is the work horse for the military and commercial user. It offers great capacity at a medium price. Although not as naturally jam resistant as the EHF spectrum, there are ways to make the SHF frequency jam resistant without adding a lot of cost. SHF is typically associated with fixed station and gateway communications. EHF is very jam resistant and mobile. EHF can operate in a scintillated environment better than UHF or SHF. Although the capacity of Milstar (the Military EHF system) is currently low, EHF, in general, offers the greatest amount of capacity in the future.

COMMUNICATION SATELLITE SYSTEMS

DoD Satellite Systems

US MILSATCOM systems operate in three specific frequency ranges primarily in geosynchronous orbits.

- The UHF spectrum in comprised primarily of UHF Follow-on (UFO) satellites and the new 3G compatible Mobile User Objective System launched in February 2012

- The Defense Satellite Communications System Phase III (DSCS III) and the Wideband Global Satellite (WGS) operate in the Super-high Frequency (SHF) spectrum.

- Milstar operates in the Extremely-high Frequency (EHF), SHF, and UHF spectrums as does the new Advanced EHF system.

Each of the three systems above provides support for a fourth system, the Air Force Satellite Communications system (AFSATCOM). AFSATCOM is not a system of dedicated satellites, but a system of dedicated channels or transponder packages riding on the satellites of the MILSATCOM system. AFSATCOM is used to disseminate Emergency Action Messages (EAMs).

Fleet Satellite Communications (FLTSATCOM)

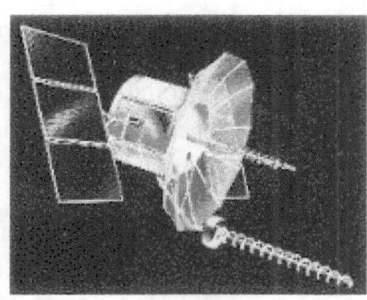

The Fleet Satellite Communications System (FLTSATCOM) was a UHF/EHF military satellite communications system. The FLTSATCOM system provided worldwide operational communications for naval aircraft, ships, submarines and ground stations. It also provided communications between the POTUS and SECDEF and the strategic nuclear forces as well as between other high-priority users. High priority users included the White House Communications Agency, reconnaissance aircraft, Air Intelligence Agency and ground forces (e.g., Special Operations Forces). However, its main purpose was for naval

Figure 6-4: Fleet Satellite Communications Satellite

afloat communications. These satellites are no longer the operational mainstay though some satellites remain in service as backup and spares for UHF communications. FLTSATCOM is worth noting here because of its outstanding service supporting the US for over two decades and there are currently two satellites still in backup/reserve.

UHF Follow-On

The UHF Follow-On (UFO) satellites replaced the Fleet Satellite Communications (FLTSATCOM) and the Hughes-built Leasat spacecraft currently supporting the Navy's global communications network, serving ships at sea and a variety of other US military fixed and mobile terminals. They are compatible with ground- and sea-based terminals already in service. The UFO satellite system is deployed as a multi-satellite constellation.

Figure 6-5: UHF Follow-On Satellite

The first UHF F/O was launched March 25, 1993. The Atlas II rocket booster malfunctioned, placing the spacecraft in a dangerously low orbit. After efforts by the 3rd Space Operations Squadron, Schriever AFB, Colo., the satellite was prevented from crashing back to Earth and was finally placed in a safe, though unusable, orbit. The second UHF F/O satellite (F-2) was launched September 3, 1993, and was successfully placed in its proper orbit, becoming the first fully operational spacecraft in a planned nine-satellite constellation. The launch of the final satellite (F-11) took place in December 2003. The satellites are arranged in pairs in four different locations above the Earth for global coverage.

Using a building-block approach, Hughes and the Navy enhanced the constellation capabilities in stages. Satellites F-1 through 3 carry UHF and SHF (super-high frequency) payloads. Beginning with F-4 the Navy added an extremely high frequency communications package. This addition included 11 EHF channels distributed between an earth coverage beam and a steerable 5-degree spot beam compatible with Milstar ground terminals. The EHF subsystem provides enhanced anti-jam telemetry, command, broadcast, and fleet interconnectivity communications, using advanced signal processing techniques. The EHF Fleet Broadcast capability supersedes the need for the SHF fleet uplink. Beginning with UHF F/O F-7, the EHF package has been enhanced to provide 20 channels through the use of advanced digital integrated circuit technology. F-7 introduced an enhancement to the EHF package that essentially doubled capacity. The SHF payload is replaced by the GBS package on F-8 through 11.

The F-1 through F-7 spacecraft include an SHF subsystem, which provides command and ranging capabilities when the satellite is on station as well as the secure uplink for Fleet Broadcast service, which is downlinked at UHF.

The UHF F/O satellites offer increased communications channel capacity over the same frequency spectrum used by previous systems. UHF F/O supports global communications to Naval forces, provides channels to replace the 5 kHz narrow-band channels previously provided by FLTSATCOM and replaces the 500 kHz DoD wide-band channel with an appropriate number of 5 and 25 kHz channels. Each spacecraft has 11 solid-state UHF amplifiers and 39 UHF channels with a total 555 kHz bandwidth. The UHF payload comprises 21 narrowband channels at 5 kHz each and 17 relay channels at 25 kHz. In comparison, FLTSATCOM offers 22 channels. Since the design is for two satellites at each orbital position, 78 UHF channels should be available over the Atlantic, Pacific and Indian Ocean regions as well as CONUS.

Each satellite has a projected orbital operational life of 14 years. The satellite is designed to operate for 30 days without ground contact if necessary. F1, F2, and F9 are no longer in service.

The Naval Space Command at Point Magu, CA took control the UFO constellation from the Air Force in July 1999. The responsibilities were transferred to the Naval Network and Space Operations Command in July 2002 which is now the Naval Satellite Operations Center. Although UFO satellites are owned by the Navy, who is also responsible for satellite communications configuration, the system will provide satellite communications to all services. Channel allocation is accomplished in the same manner as for FLTSATCOM. The JCS has mandated that all UHF SATCOM radios operate in the Demand Assigned Multiple Access (DAMA) mode unless a waiver has been granted. DAMA is a modified time sharing technique to allow more users to share the same UHF channel, 5 kHz or 25 kHz.

Global Broadcast System (GBS)

In order to meet the demands of a rapidly deployed, highly mobile force structure, today's in-the-field warfighter demands a high data rate information infrastructure. The Global Broadcast Service capitalizes on the popular commercial direct broadcast satellite

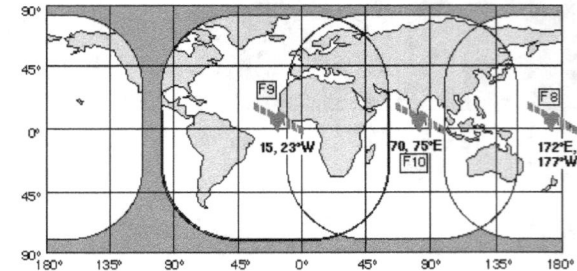

120

Figure 6-6: Location of GBS Satellites

technology to provide critical information to the nation's warfighters. The GBS system is a space based, high data rate communications link for the asymmetric flow of information from the US or rear echelon locations to deployed forces. It is designed to provide information in a dynamically reconfigurable format rapidly adaptable to peace and wartime circumstances, and deliver it to theaters of operation worldwide.

GBS is an extension of the Defense Information Systems Network (DISN) and a part of the overall DoD MILSATCOM Architecture. As such, it is designed to employ an open architecture which can accept a variety of input formats. It exploits commercial off-the-shelf (COTS) technology. It must interface with, and augment other major DoD information systems, such as the Global Command and Control System (GCCS), as well as other theater information management systems. Eventually, GBS may supplant some theater information management systems.

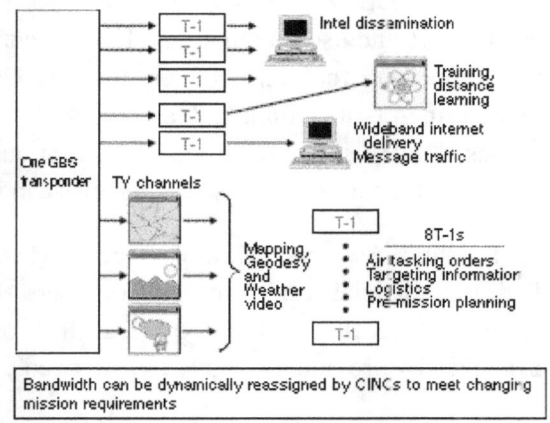

The first GBS payload was put into service in 1998 on UFO F-8. UFO F-9 was launched in October 1998; however it is no longer operational for GBS. Full three-satellite operational capability was gained with the launch of F-10 (see **Figure 6-6**). GBS is currently hosted as a payload on the Wideband Global Satellite (WGS) system.

Figure 6-7: Types of GBS Services

The GBS payload replaces the SHF X-band payload with four 130-watt, 24 mega-bits-per-second (Mbps) military Ka-band (30/20 GHz) transponders. There are three steerable downlink spot beam antennas (two at 500 nm and one at 2000 nm) as well as one steerable and one fixed uplink antenna. This modification results in a 96 Mbps capability per satellite. The system will transmit to small, mobile, tactical terminals. **Figure 6-7** shows the types of services that can be provided by one 24 Mbps spacecraft transponder, a vast increase over today's warfighter capability.

High-power satellite transponders, which provide high-speed, wideband, simplex broadcast signals, characterize the Phase II Global Broadcast Service (GBS). This information is disseminated to small, 22-inch-diameter, mobile, and affordable tactical terminals. Broadcast management centers provide the information management to package, schedule, and deliver the broadcast product. They also respond to user information requests from the field. Typical information products include video, mapping, charting and geodesy, imagery, weather, and digital data.

Data is received by the satellite via a fixed receive antenna from a broadcast management center (primary injection point) and a steerable receive antenna from theater injection point(s). Each of the four transponders can be accessed through either of the receive paths, configured by ground command. Data is transmitted on three steerable spot beam antennas per spacecraft into

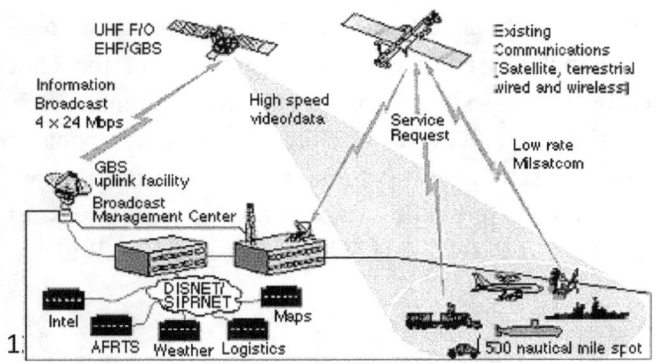

Figure 6-8: How GBS Works

22-inch ground receive antennas. Each of two spot beams covers an area of 500 nautical miles in diameter at the sub-satellite point and supports data rates of up to 24 Mbps per transponder, with two transponders assigned to each of the spot beams. The third downlink spot beam covers an area of 2,000 nm in diameter at the sub-satellite point, supporting a data rate down to 1.5 Mbps. One of the transponders is switchable by ground command from the 500 to the 2000 nm spot beam.

Demand Assigned Multiple Access (DAMA)

The evolution of UHF SATCOM has seen enormous changes in mission philosophy, technical capabilities, and user population. The current theater commander wants high speed encrypted voice and data supporting thousands of terminals. UHF SATCOM is a highly flexible and low cost means for beyond-line-of-site command and control. As tactical satellite communications requirements have increased, UHF satellite capacity has been quickly saturated. The available frequencies are limited and only the efficient use of those frequencies can satisfy the increasing demands of the warfighter.

Demand Assigned Multiple Access (DAMA) is automated channel sharing. Instead of having dedicated channels and transponders, users share the capacity. The unused transponder space can be dynamically reallocated in near real-time on the basis of precedence. This increases the loading efficiency by providing roughly four to twenty times the information throughput of the current systems.

The DAMA Control Station divides a channel into segments of time called "time slots." A user terminal interacts with the control station, which dynamically allocates time slots for that user's communications. Channel resources are allocated on the basis of current needs and network rankings. Any unused DAMA channel resources are available to be shared by everyone.

Mobile User Objective System (MUOS)

The Mobile User Objective System (MUOS) is the next generation narrowband system for the mobile warfighter and is the successor to UFO. The constellation will consist of four operational satellites with one on-orbit spare with the first launch scheduled for February 2012. MUOS segments includes the new satellites, ground control systems, ground terminals, and gateways/teleports.

Ground terminals will migrate from the AN/PSC-5, Spitfire to the SCAMP and then eventually to the Joint Tactical Radio System (JTRS). The JTRS will be a software programmable and modular communications system that will be interoperable with legacy waveforms. When JTRS is fully fielded, true interoperability will then finally be achieved between Services.

Defense Satellite Communications System Phase III (DSCS III)

DSCS III was designed to provide SHF wideband communications for worldwide long haul communications to fixed and mobile national, strategic, tactical and other designated governmental users. This includes Presidential communications, the Global Command and Control System (GCCS), from early warning sites to operations centers, and Combatant Commands and tactical forces. The system consists of DSCS III satellites in geosynchronous orbit, ground control stations and user terminals. Residual satellites with some limited operational capability are available for increased traffic capacity and flexibility in providing coverage to specific areas around the world. The final DSCS III satellite was launched in 2004 and is being replaced by the Wideband Global SATCOM (WGS) satellite.

Under normal operating conditions DSCS III provides substantial worldwide capacity of high quality voice circuits and wideband data circuits. During and after nuclear attack or when being jammed, the system concentrates its capabilities by reducing the number of channels to meet minimum essential communications requirements of high priority users. The US Army Space and Missile Defense Command (SMDC) performs DSCS III communications payload control while Air Force Space Command (AFSPC) provides command and control of the satellite vehicles in the constellation via the 50th Space Wing's 3rd Space Operations Squadron.

The first DSCS III satellite was launched in October 1982. There are currently ten DSCS III satellites in geostationary orbit, five of which are primary and five are in reserve. Each of the five operational satellites and spare satellites has a primary and alternate network control station located at major nodes such as Ft. Detrick, Maryland. The design life for each satellite is 10 years. The satellites are at an altitude of approximately 22,300 miles in a geostationary orbit around the equator. All ten satellites are in continuous 24-hour operations with the reserves/spares primarily used for Ground Military Force (GMF) training missions.

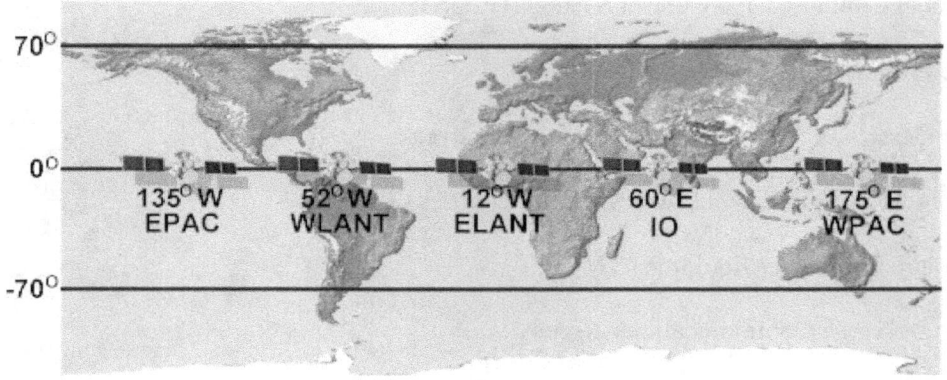

Figure 6-9: Primary DSCS Locations

The primary DSCS III satellites provide overlapping footprints for worldwide communications between 70° North latitude and 70° South latitude. Communications beyond these latitudes becomes very weak due to earth's flattening above 70° latitude to the poles. Heavy terminals, such as the FSC-78 with the large 60 foot antenna, could access a DSCS III satellite from some locations above 70° North or below 70° South latitude. The ten satellite constellation of DSCS III allows most Earth terminal locations to access at least two satellites.

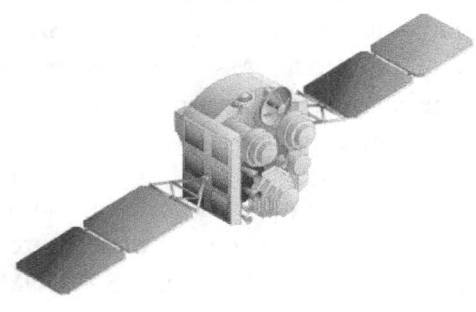

The DSCS III frequency plan falls within the SHF spectrum (X band) with uplink frequencies of 7900MHz to 8400 MHz which the transponders down-translate to the downlink frequencies of 7250 MHz to 7750 MHz. Any type of modulation or multiple access may be used since none of the transponders process or demodulate the signals.

There are two communications subsystems on the DSCS III. The primary system has eight antennas which can be

Figure 6-10: DSCS III

connected in various ways to six independent transponders. Each transponder has its own limiter, mixer and transmitter so that it can be configured to serve specific types of user requirements. There are two earth coverage horns and one multi-beam receiving antenna. The multi-beam antenna can form any beam of arbitrary size and shape, and can be oriented toward specific locations within the footprint by means of a beam-forming network that controls the relative amplitudes and phases of each of 61 individual beams. This antenna can also form "nulls" in selected directions to counter jammers on the ground. Two transmitters on the satellite are always connected to earth coverage antennas. One antenna is a gimbaled fish antenna which provides a 3-degree spot beam. The other earth coverage antennas are two 19-beam transmit multi-beam antennas. These antennas do not have the "nulling" capability.

The secondary communications subsystem on DSCS III is AFSATCOM. This package has its own UHF transmitting and receiving antennas, but can be connected to the earth coverage or multi-beam receiving antennas.

DSCS III provides range extension for the following types of networks:

- Global Command and Control System (GCCS)

- Defense Switched Network (DSN)

- Jam Resistant Secure Communications Networks

- Tactical Warning/Attack Assessment Networks (TW/AA)

- Mobile Subscriber Equipment (MSE)

- White House Communications Agency

- Navy Flagship Command and Control Network

- Ground Mobile Forces and AFLOAT communications

Ground terminal characteristics such as transmit power, antenna size, and antenna elevation angle with respect to the satellite determine a terminal's ability to communicate within a footprint from its particular location.

Satellite Control

The Chairman, Joint Chiefs of Staff has primary responsibility for DSCS III with the Defense Information Systems Agency (DISA) having management responsibility for its control segment. DISA is a DoD agency that reports directly to the Chairman of Joint Chiefs of Staff and the Assistant Secretary of Defense for C3I. The DISA mission is to develop, test, manage, acquire, implement, operate and maintain information systems for C4I and mission support under all conditions of peace and war.

USARSTRAT/SMDC's, (1st Space Brigade) 53rd Signal Battalion (Satellite Control), has a critical role in the worldwide operation of DSCS III. Its mission is to provide communications network control for the DSCS III. The 53rd Signal Battalion operates and maintains the five Wideband SATCOM

Operations Centers (WSOCs) worldwide. The WSOCs provide real-time monitoring and control for the DSCS III, WGS, and GMF networks. They provide payload control of the satellites, which involves commanding changes to transponder and antenna configuration.

The control segment allocates satellite capacity to best serve user requirements. Control segment computer algorithms provide an allocation process that makes use of the considerable flexibility of the DSCS III and WGS satellites. The control segment optimizes the network configuration, responds to jammers and generates command sets to configure the satellite and processes telemetry from the satellites.

To accomplish these responsibilities, the US Army SMDC operates the WSOC's, Regional Space Support Centers (RSSC), the AN/MSQ-114 SATCOM Control Terminals, and provides personnel in the DSCS III Control Facility which is part of the Consolidated Space Operations Center (CSOC) at Schriever Air Force Base, Colorado.

Wideband SATCOM Operations Center (WSOC)

The WSOCs perform payload control on DSCS III satellites and control the user network. Each WSOC is a control center for a designated DSCS III or WGS. There are currently WSOC's in operation at Fort Meade, Maryland; Fort Detrick, Maryland; Wahiawa Annex, Hawaii; Landstuhl, Germany and Fort Buckner, Japan.

Regional Space Support Centers

Regional Space Support Centers (RSSC) are another component of the US Army SMDC. There are three RSSCs: Washington, D.C.; Wheeler Air Force Base, Hawaii; and Vaihingen, Germany. The RSSCs are focal points for Combatant Commander tactical MILSATCOM requirements. At present the focus of the RSSCs' is on DSCS III; however, there are concepts to expand the RSSCs role to handle requirements for other MILSATCOM systems, DMSP, GOES, NOAA satellites, GPS, and surveillance and warning systems.

DSCS III Access

Access to the DSCS III satellites is accomplished differently for the DSCS III network and the GMF network. For DSCS III network access, the following is a summary of the process:

- Users identify their requirements.

- Users submit their requirements to their supporting Combatant Commander.

- The joint command J6 will coordinate with DISA (GCC or RCC) for the required resources.

- DISA will engineer the link parameters to support the requirements. The information is passed to the DSCS III Ops Centers where the Network Controllers add/subtract/ monitor the entire net.

- The user is informed of the circuit design (power/bandwidth/times of usage).

- Communication stays open between all parties to assure the warfighters' needs are met.

For GMF access, the tactical user receives mission tasking and begins the planning process with the Communications Systems Planning Element (CSPE). The CSPE determines the mission's satellite communications requirements and develops a Satellite Access Request (SAR) for the RSSC. The SAR consists of the following:

- Who, When, What, Where and How

- Unit and Mission, date/time, data rate, terminal types and location, network configuration and priority

- The RSSC (Co-located with DISA-GCC or RCC) will:

- Coordinate with DISA for resources to support the SAR

- Perform network planning with parameters given by DISA if the SAR can be supported.

- Develop Satellite Access Authorization (SAA) with the satellite, look angles, power, frequency and controller.

The SAA is sent to the originating CSPE, DISA and the controller. The CSPE produces deployment orders and configuration sheets for terminals while the DISA directs the controlling DSCSOC to update their operational database. Finally, 30 minutes prior to the mission start time, the controller contacts the terminals and directs access to the satellite.

Wideband Global Satellite (WGS) System

In 2000, the Space and Missile Systems Center (SMC) led a multi-service program to acquire a new series of communications. One WGS satellite has more capacity than the entire constellation of DSCS III satellites. The first WGS was launched on October 10, 2007.

The WGS space segment will consist of at least three satellites in geosynchronous orbit. Each satellite will provide coverage and capacity to military forces operating anywhere within a field of view from at least 65 degrees North to at least 65 degrees South latitude.

Through nine X-band antennas and ten Ka-band antennas, each satellite will provide two-way X-band, two-way Ka-band and broadcast Ka-band services. The X-band connectivity will be fully compatible with the existing DSCS III service and the Ka-band broadcast capability will be fully compatible with existing GBS service.

Military two-way Ka-band services and crossbanding (X- to Ka- or Ka- to X-) is a new capability initiated on the WGS.

Each satellite will have the flexibility and capability to provide the coverage and capacity needed to support military forces operating in a Major Theater War and/or Small Scale Contingencies. Each satellite will be able to focus its resources to support high concentrations of users in very small areas, surrounding users operating within the theater, and dispersed users transiting to a theater or performing other military missions. The satellites will also be able to provide support to naval, ground, and air forces performing their day-to-day global missions.

As the WGS satellites are launched, current X-band terminals will capable of operating on either DSCS III or WGS. Additionally, WGS will allow GBS information to be transmitted X-band, crossbanded to Ka-band, and received on existing GBS receive suites. There are no existing terminals that use the military Ka-band required for the two-way Ka-band service; however, the Army will develop ten Engineering Developmental Models to operate over WGS.

Milstar

Milstar (which stands for Military Strategic, Tactical and Relay) satellites operate primarily in the Extremely High Frequency (EHF) and Super High Frequency (SHF) bands. Milstar satisfies the US military's communications requirements with

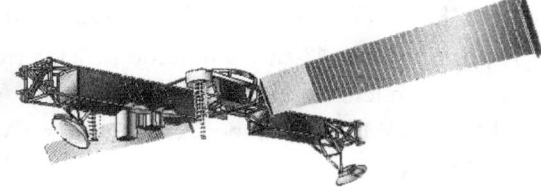

Figure 6-11: Milstar

worldwide, anti-jam, scintillation resistant, Low Probability of Intercept (LPI) and Low Probability of Detection (LPD) communications services.

Milstar is designed to meet the minimum essential command, control and communications requirements of our POTUS and SECDEF and strategic and tactical multi-service military forces well into the 21st century. The system will allow flexible reconfiguration of the transponders and antennas to optimize the allocation of resources in the satellites. The satellites are in low-inclination geosynchronous orbit at an altitude of approximately 22,300 miles. This orbit gives coverage from 65 degrees north to 65 degrees south. Survivability and durability requirements are satisfied by anti-jam, hardening and system autonomy features.

There are two types of satellites. The Milstar I satellites carry a secure, robust low-data-rate (LDR) communications payload, and a crosslink payload that allows the satellites to communicate globally without using a ground station. The Milstar II satellites extend the communications capabilities to higher data rates by adding a medium-data-rate (MDR) payload. The Milstar I and II satellites are fully interoperable for LDR communications and crosslinks. Using the crosslinks, the constellation supports multiple users simultaneously without reliance upon physically vulnerable nodes. Milstar satellites provide a "communications switch in the sky."

Milstar I satellites carry a low data rate payload that provides worldwide, survivable, highly jam-resistant communications for the POTUS and SECDEF as well as tactical and strategic forces. Advanced processing techniques on board the spacecraft as well as satellite-to-satellite cross linking allow Milstar satellites to be relatively independent of ground relay stations and ground distribution networks. The first Milstar I was successfully launched on 7 February 1994, and the second, on 6 November 1995.

In October 1993, SMC awarded a contract for development of the Milstar II satellite, which carried both low and medium data rate payloads. The addition of the medium data rate payload greatly increased the ability of tactical forces to communicate within and across theater boundaries. Only four Milstar II satellites were produced because DoD decided in 1993 that they were to be replenished by a new, lighter, cheaper series of Advanced EHF satellites. Unfortunately, the first Milstar II satellite went into an unusable orbit on 30 April 1999. The next two Milstar II satellites were successfully launched on 27 February 2001 and 16 January 2002 to complete an on-orbit constellation of four satellites. The sixth and final Milstar satellite was successfully launched on 8 April 2003.

LDR Payload

The LDR payload offers nearly 200 user channels and relays coded teletype and voice messages at data rates of 75 to 2400 bits per second. The first two satellites are block one birds with only a Low Data Rate (LDR) (75 to 2400 BPS) capability. Milstar flight 1 is positioned at 120° west longitude and flight 2 is positioned at 4° east longitude Milstar LDR supports strategic and tactical requirements for high anti-jam and nuclear scintillation protection. The strategic users are the same as those currently using AFSATCOM. To support this mission, Milstar is designed to be a far more survivable system than AFSATCOM.

MDR Payload

The MDR payload provides secure, jam-resistant communications services through unique onboard signal and data processing capabilities. It sends real-time voice, video and data to military personnel in the field at rates up to 1.5 Mbps. The payload uses a 32-channel EHF uplink and a SHF downlink. The MDR payload dynamically sorts incoming data and routes them to the proper downlinks to establish networks and provide bandwidth on demand. If necessary it passes the data on to another satellite via crosslink.

The MDR antenna coverage subsystem consists of eight narrow spot beam antennas provided by TRW: two narrow spot beams with nulling capabilities (nuller antennas) and six distributed user coverage antennas, each supporting two-way communications. In contrast to commercial communications satellites, whose beams can cover entire continents, Milstar's beams are very narrow, providing less opportunity for enemy detection and penetration. The nuller antennas resist jamming from within their respective coverage areas by changing their gain patterns when a jamming signal is detected. The distributed user coverage antennas provide high gain/low side lobes for distributed users.

Crosslink Payload

Like a handshake in space, crosslinks provide rapid, secure communications by enabling the satellites to pass signals to one another directly through space while requiring only one ground station on friendly soil. The crosslink payload provides V-band (60 GHz) data communications between Milstar satellites for both the MDR and LDR payloads. This includes modulation and demodulation of the data, upconversion, amplification for transmission and downconversion.

	MDR	LDR
Data rate	4.8 kbps - 1544 kbps	75 bps - 2400 bps
No. communications channels	32	192
No. users/channel	1-70	1-4
Coverage	8 high-gain narrow spot beams	2 narrow spot beams 1 wide spot beam 1 earth coverage antenna 1 UHF transmit antenna 1 UHF receive antenna 5 earth coverage uplink agile beams 1 earth coverage downlink agile beam
Anti-jam	Waveform Nuller - active Transmission security	Waveform Transmission security

Table 6-1: Milstar Features

Milstar was the world's first satellite constellation, government or commercial, to employ crosslinks. A message was uplinked from the national Military Command Center at Fort Belvoir, VA., through the Milstar F-1 satellite. It was crosslinked to the Milstar F-2 spacecraft, then downlinked to commanders at Pacific Command at Camp H.M. Smith, Hawaii, and US Atlantic Command at Norfolk, VA.

The Milstar system provides uplink communications at extremely high frequency (EHF), 44 GHz, and ultrahigh frequency (UHF), 300 MHz, and downlink communications at super-high frequency (SHF), 20 GHz, and UHF, 250 MHz. The crosslinks operate in the 60 GHz region.

Communications features of the MDR and LDR payloads are compared in Table 5-1 above.

Advanced EHF System

The Advanced Extremely High Frequency (AEHF) System is a joint service satellite communications system that will provides near-worldwide, secure, survivable, and jam-resistant communications for high-priority military ground, sea, and air assets. In view of the limited future of the Milstar system, SMC began the acquisition of this follow-on EHF military communications system, known ultimately as AEHF.

The system will consist of four satellites in geosynchronous earth orbit (GEO) providing up to 100 times the capacity of the 1990s-era Milstar satellites, servicing up to 4,000 networks and 6,000 terminals. This constellation will provide continuous 24-hour coverage between 65 degrees north and 65 degrees south latitude. Advanced EHF will allow the National Security Council and Unified Combat Commanders to contact their tactical and strategic forces at all levels of conflict through general nuclear war and supports the attainment of information superiority. AEHF will provide connectivity across the spectrum of mission areas, including land, air, and naval warfare; special operations; strategic nuclear operations; strategic defense; theater missile defense; and space operations and intelligence. The AEHF System will provide warfighters with broadcasting, data networking, voice conferencing, and strategic report-back capabilities. It will also provide commanders with the advantages of near-worldwide coverage, multi-user connectivity, protected data, and ease of use. AEHF protections include anti-jam capabilities, Low Probability of Detection (LPD), a Low Probability of Intercept (LPI), and advanced encryption systems. Finally, the AEHF system is a multinational effort with international partners from the United Kingdom, the Netherlands, and Canada. These international partners will gain access to the AEHF network through their own terminals.

The AEHF system will consist of three segments: space (the satellites), terminals (the users), and mission control and associated communications links. The segments will provide communications in a specified set of data rates from 75 bps to approximately 8 Mbps. The space segment consists of a cross-linked constellation of satellites to provide worldwide coverage. The mission control segment controls satellites on orbit, monitors satellite health, and provide communications system planning and monitoring. This segment is highly survivable, with both fixed and mobile control stations. System uplinks and crosslinks will operate at extremely high frequency (EHF), and downlinks at super high frequency (SHF). The terminal segment includes fixed and mobile ground terminals, ship and submarine terminals, and airborne terminals. User terminals supported by AEHF include Secure Mobile Anti-Jam Reliable Tactical-Terminal (SMART-T), Single Channel Anti-Jam Man Portable (SCAMP), Family of Advanced Beyond Line-of-sight Terminals (FAB-T), and Navy Multiband Terminals (NMT). The AEHF satellites will respond directly to service requests from operational commanders and user terminals providing real-time point-to-point connectivity and network services on a priority basis. On-board signal processing will provide protection and ensure optimum resource utilization and system flexibility among the Military Services and other users who operate terminals on land, sea, and air. The AEHF system will be backward compatible with the low data rate (LDR) and medium data rate (MDR) capabilities of legacy Milstar satellites and terminals, while providing extended data rates (XDR) and other improved functionality at substantially less cost than the previous system. Each satellite will be launched with the Evolved Expendable Launch Vehicle (EELV); the initial launch is planned for 2008-2010. The MILSATCOM Joint Program Office (MJPO) of SMC is responsible for development, acquisition, and sustainment of the AEHF Program.

The system will be compatible with Milstar elements and would incorporate them throughout their useful lifetimes. Like Milstar, but greatly enhanced, the AEHF system will feature on-board signal processing and satellite crosslinks to eliminate reliance on ground stations for routing data. Data uplinks to the satellites and crosslinks between satellites will operate at EHF, and downlinks will operate at SHF.

Air Force Satellite Communications System (AFSATCOM)

There are no separate AFSATCOM satellites. Instead, AFSATCOM transponders are packages added to other satellite systems. AFSATCOM provides secure, reliable and survivable two-way global communications between the POTUS and SECDEF and the strategic nuclear forces. The AFSATCOM system is used for Emergency Action Message (EAM) dissemination, JCS/Combatant Commander Internetting, Combatant Commander force direction message dissemination, force report back and other high-priority user traffic dissemination. Strategic nuclear forces include ICBM launch and control centers, B-52, B-1B and B-2 bombers and nuclear capable submarines (SSBNs).

Primary satellite systems in geosynchronous orbit carrying AFSATCOM packages include DSCS III and Milstar. The DSCS III payload offers an alternate path for EAM dissemination. There are two systems in use for polar coverage: the Satellite Data System (SDS) and Package D, a piggyback payload on classified host vehicles. SDS satellites include a payload similar to the twelve-channel 5 kHz system that was onboard the FLTSATs. However, all twelve are regenerative and can only be used for 75 BPS data. Package D satellites provide a UHF package similar to the SDS satellites.

There are twelve AFSATCOM 5 kHz channels independent of other communication systems on their host satellites. Seven of the twelve 5 kHz narrow-band channels are regenerative and can only be used for 75 BPS digital communications (not voice). Additionally, the CONUS, Atlantic and Pacific AFSATCOMs have a 500 kHz UHF wideband transponder. This transponder provides connectivity for the JCS/COMBATANT COMMANDER Internets which frequency-hop within the 500 kHz bandwidth or it can be divided into 21 accesses of 25 kHz allowing customers to share the 500 kHz bandwidth.

Channel capacity is allocated by the JCS. AFSATCOM is "owned" by the Air Force.

Other Military SATCOM Systems

There are far too many satellite systems to discuss in this text. Communications is vital to every country. Satellite communications gives each country the broadcast and range extention capbilities to contect even the most remote areas. Therefore, each country typically places importance on communications satellites as the first satellite priority. The SATCOM can be owned by the country it supports or a leased communications package on another country's satellite. There are several cases where two or more countries form a consortium to build satellites that will support more than one country.

This text will only discuss the foreign satellites that are used to support international military operations. It will not discuss threat SATCOM.

SKYNET

SKYNET 4 first entered service in 1988. Three satellites were built by British Aerospace and Marconi (now part of Matra Marconi Space (MMS)) for UK service and two more for NATO (NATO IV). These satellites are controlled on behalf of the three UK armed services and NATO by the Royal Air

Force (RAF) from ground stations in the UK. Skynet 4B and Skynet 4C were launched by Arianespace in December 1988. As the existing UK satellites (SKYNET 4A, B and C) reach the end of their operational lives, after nearly a decade's service, they were replaced by three further satellites (SKYNET 4D, E and F), known collectively as SKYNET 4 Stage 2. SKYNET 4D was launched in January, 1998 to replace 4B. 4E was launched in February, 1999. The final satellite of the program, SKYNET 4F, was been launched in 2001.

The newest SKYNET 5 series satellites have an enhanced communications package that provides UHF (Ultra High Frequency) and SHF (Super-High Frequency) communications services designed to support the UK armed forces in their enhanced roles, such as the NATO Rapid Reaction Force and support of humanitarian aid anywhere in the satellite's coverage. Multiple beams with nulling capabilities at SHF, together with the EHF payload, can provide improved capacity, performance and services combined with ease of use. SKYNET 5A and 5B are currently in service.

A Memorandum of Understanding was signed with France and Germany to undertake a collaborative Project Definition phase for this program, under the title of TRIMILSATCOM.

NATO SATCOM Post-2000

The North Atlantic Treat Organization (NATO) had a constellation of NATO III and IV satellites providing a "general purpose" military communications system. These SHF satellites operated in the military X-band were designed to provide communications and intelligence support to the Combatant Commanders, NATO and to the national command authority of NATO forces. US forces in their NATO role use the NATO constellation.

The Satcom Post-2000 program gives the Alliance improved satellite communication capabilities, which is important as NATO forces take on expeditionary missions far beyond the Alliance 's traditional area of operations.

Under the program, the British, French and Italian governments are providing NATO, through what is known as "Capability Provision", with advanced satellite communication capabilities for 15 years as of January 2005.

The benefits include increased bandwith, coverage and expanded capacity for communications and data, including with ships at sea, air assets, and troops deployed across the globe.

The program provides NATO with access to the military segment of three national satellite communications systems - the French Syracuse, Italian Sistema Italiano di Communicazioni Riservate ed Alarmi (Italian Classifed and Emergency Communications System, abbreviated SICRAL), and British Skynet 4/5 - under a Memorandum of Understanding.

This new system replaces the two existing NATO IV communications satellites, which were launched in 1991 and 1993, respectively, and were designed to last 10 years.

The Satcom Post-200 program gives NATO access to two components: super high frequency (SHF) and ultra high frequency (UHF) communications. UHF (300-3,000 MHz) is used for tactical communications, SHF (3-300 GHz) for ground stations with larger radar dishes.

All three systems (Syracuse, SICRAL, and Skynet 4/5) can provide SHF communications, while SICRAL and Skynet provide UHF communications.

Commercial and Civil SATCOM Systems

The warfighter will always require the ability to communicate over DoD satellite systems. However, these systems will never satisfy all of the requirements for SATCOM to support military operations. Several studies have clearly shown that many existing and emerging narrowband requirements could be satisfied by commercial SATCOM systems. The off-loading of those requirements would help in relief of UHF SATCOM for the Army. The US military is actively involved with the civil and commercial SATCOM industry in identifying current and emerging technologies that can be leveraged to support the military's goal of full-spectrum dominance. The rapid product cycle that the commercial industry can generate is something that DoD can rely upon and leverage to upgrade military battlefield systems.

The most critical communications in hostile threat environments will be placed on military SATCOM systems. The administrative and logistics traffic as well as peacetime operations traffic can be satisfied more flexibly and economically quite possibly by commercial means. The Army is already using commercial SATCOM extensively (INMARSAT and Iridium represent two of the better known systems). Commercial leases of C- and Ku-band capacity have long been accepted as part of the MILSATCOM system.

Tactical applications for commercial satellites have been tested successfully during recent military operations around the world. Commercial satellites not only provide interoperability between services but also between allied nations. US peacekeeping forces in Bosnia and Kosovo currently use a commercial telecommunications system provided by Spring, Inc. which provides voice and data access to local, worldwide long-distance, and internet services via SATCOM. Alascom transportable terminals, initially deployed to support training exercises in Alaska, were transferred to use in Panama during Operation Just Cause. PanAmSat, another commercial satellite system, has provided satellite links to the Army in support of drug interdiction programs in Bolivia and Peru. In Operations DESERT STORM and DESERT SHIELD, INTELSAT and INMARSAT were used extensively. Forces in Operations IRAQI FREEDOM and ENDURING FREEDOM use Iridium and Thuriya communications systems.

Commercial satellites have limitations, however, that must be mitigated by careful planning. See the section on Planning Considerations to compare the differences between military and commercial satellite systems.

This section will cover some of the more extensively used systems. This is not intended to be all encompassing.

International Telecommunications Satellite (INTELSAT)

The foremost international commercial provider of fixed-site satellite communications services is INTELSAT. INTELSAT was created in 1964, formed on the basis of an international treaty signed by US President John F. Kennedy the previous year. It was a UN sponsored, not-for-profit, commercial consortium comprising 143 member countries and signatories, but in 2001 it became a private company.

Intelsat was sold for $3.1bn in January 2005 to four private equity firms. The company then acquired PanAmSat on July 3, 2006, and is now the world's largest provider of fixed satellite services, operating a fleet of 53 satellites in prime orbital locations. Intelsat maintains its corporate headquarters in Bermuda, with a majority of staff and satellite functions — administrative headquarters — located at the Intelsat Corporation offices in Washington, DC. A highly international business, Intelsat sources the majority of its revenue from non-US located customers.

Spacecraft operations are controlled through ground stations in Clarksburg, Maryland (USA), Hagerstown, Maryland (USA), Riverside, California (USA), and Fuchsstadt, Germany.

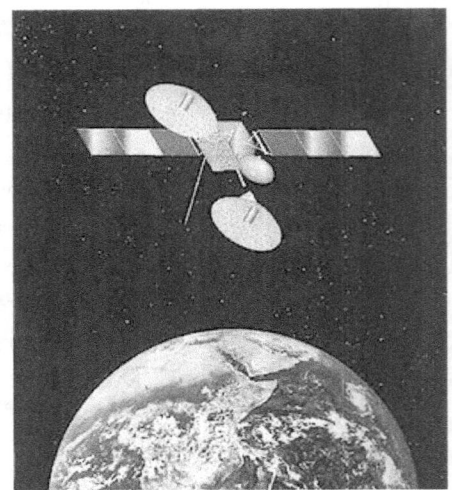

Figure 6-12: INTELSAT

Tracking and Data Relay Satellite System (TDRSS)

The Tracking and Data Relay Satellite System (TDRSS) is a communication signal relay system which provides tracking and data acquisition services and serves as the sole means of continuous, high-data-rate communication with the Space Shuttle, with the Space Station upon its completion, and with dozens of satellites in low earth orbit. The system is capable of transmitting to and receiving data from customer spacecrafts over at least 85% of the customer's orbit. LANDSAT is an example of a satellite system in low earth orbit that uses TDRSS to relay information.

The TDRSS space segment consists of six on-orbit satellites located in geosynchronous orbit. Three TDRSs are available for operational support at any given time. The operational spacecraft are located at 41, 174 and 275 degrees west longitude. The other TDRSs in the constellation provide ready backup in the event of a failure to an operational spacecraft and, in some specialized cases, resources for target of opportunity activities.

The steerable, single-access antennas can simultaneously transmit and receive at S-band and either Ku- or Ka-band, supporting dual independent two-way communication. The selection of Ku- or Ka-band communications is done on the ground. Receive data rates are 300 megabits/second at Ku- and Ka-band, and 6 Mbps at S-band. The spacecraft carries additional capability for Ka-band receive rates of up to 800 Mbps. Transmit data rates are 25 Mbps for Ku- and Ka-band, and 300 kilobits/second for S-band. In addition, an S-band phased array antenna can receive signals from five spacecraft at once, while transmitting to one.

International Maritime Satellite (INMARSAT)

INMARSAT is an internationally owned satellite consortium that provides mobile satellite communications services to the shipping, aviation, offshore, and land mobile industries. INMARSAT, with its headquarters in London, England, has 86 member nations.

INMARSAT was originally created in 1979 to serve a global maritime industry by developing a communications satellite system for distress and safety applications. INMARST has expanded significantly to include services to land mobile and aeronautical user terminals.

The system uses 10 operational satellites in geostationary orbit. Currently there are three INMARSAT-2, five INMARSAT-3, and two INMARSAT-4 satellites. The communications payload operates in the L-band and C-band. Each INMARSAT-3 and 4 carries a navigation transponder designed to enhance the accuracy, availability and integrity of the GPS and GLONASS satellite navigation systems.

The use of INMARSAT in the military has grown since Operation DESERT SHIELD and DESERT STORM when it was used extensively. There are now over 640 INMARSAT terminals on hand throughout the Army. However, the INMARSAT organization is very specific about how their system can be used. Army users are restricted from accessing INMARSAT resource except under the limitation prescribed by the INMARSAT international consortium charter which states it be used exclusively "for peaceful purposes." Basically, INMARSAT may be used by the military in life endangering situations, UN peacemaking/peacekeeping operations and forces not involved in armed conflict. The Army policy adheres to these guidelines.

Iridium

Iridium is a satellite based personal communications system from Motorola intended to allow service from any point on the globe. Originally consisting of 77 satellites (the atomic number of iridium, hence the name), the system has been redesigned to use only 66 low earth orbiting satellites. Starting on May 5, 1997, the entire constellation was deployed within twelve months on launch vehicles from three continents: the US Delta II, the Russian Proton, and the Chinese Long March.

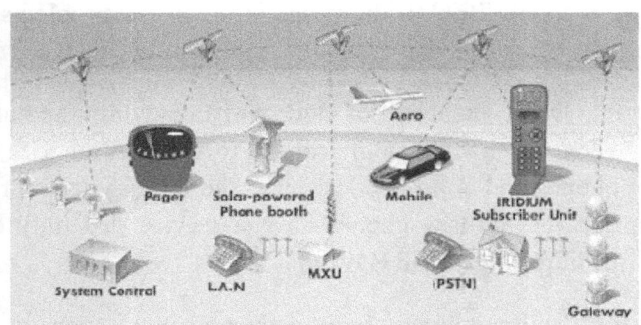

Figure 6-13: How Iridium Works

Iridium is designed to provide commercial voice, data, paging, facsimile, and messaging services. It allows for the use of handheld units much like those used in the cellular industry. Iridium is the first full LEO system in operational orbit to be used by DoD personnel and it is currently under sole source contract to provide DoD Mobile Satellite Service (MSS).

The Iridium constellation is at an altitude of about 420 nautical miles. The satellites are in near-polar circular orbits inclined at 86.4□ and distributed into six planes separated by 31.6□ around the equator with eleven satellites per plane. There is also one spare satellite in each plane. This low constellation dramatically cuts down the propagation delay of transmissions. It allows for the use of lower powered units. The biggest disadvantage of such a low orbit is the large number of satellites required for global coverage because the satellites pass over the earth quickly. The period of revolution of an Iridium satellite is approximately 100 minutes, so that any given satellite is in view overhead of a user for about 9 minutes at a time.

The system employs L-Band using FDMA/ TDMA to provide voice at 4.8 kbps and data at 2400 bps with 16 dB margin. Each satellite has 48 spot beams for Earth coverage with each beam covering an area the size of the state of Arizona. The satellite has three main beam phased array antennas, each of which serves 16 cells. Each satellite has a capacity of about 1100 channels. However, the actual number of users within a satellite coverage area will vary and the distribution of traffic among cells is not symmetrical. A key feature of Iridium is the use of crosslinks to route traffic directly to other satellites. For the warfighter this means less delay in transmissions and it is also a hedge against jamming and interception of traffic by enemy forces. Iridium uses Ka-Band for crosslinks and ground commanding.

The master control facility for the Iridium constellation is located outside of Washington, DC in northern Virginia. With the aid of three TTAC's (Telemetry, Tracking, and Control centers) located in Canada and Hawaii, this facility will regulate the positioning of the satellites during the initial placement and the ensuing orbit. The system is coordinated by 12 physical gateways distributed around the world, although in principle only a single gateway would be required for complete global coverage. The potential for Iridium's use by the warfighter was considered so great that DoD invested over 14 million dollars to build and have exclusive use of an Iridium gateway in Hawaii.

In late 1999, Iridium ran into financial problems and declared bankruptcy. A consortium of investors purchased the entire Iridium system and made it operational again.

DISA, on behalf of DoD, had already invested in the establishment of the military gateway in Hawaii. DISA contracted with Iridium to provide the US military with approximately 5,000 Iridium portable phones and unlimited service. The military Iridium phones are different from civilian Iridium phones in that the military phones have a service that only downlinks from the satellites to the DISA operated Iridium gateway in Hawaii.

Iridium phones are used extensively in Afghanistan and Iraq, and in many other parts of the world.

Globalstar

Globalstar is a consortium of leading international telecommunications companies originally established in 1991. It is a wholesale provider of mobile and fixed satellite-based telecommunications system designed to provide voice, messaging, roaming, and position location services. As Globalstar expands its services it will also offer a range of data services and facsimile services. Globalstar transmits calls from your wireless phone or fixed phone station to a terrestrial gateway, where they are passed on to existing fixed and cellular telephone networks in more than 100 countries on 6 continents.

Globalstar began its progressive roll out of service in September 1999. To ensure quality service to its users,

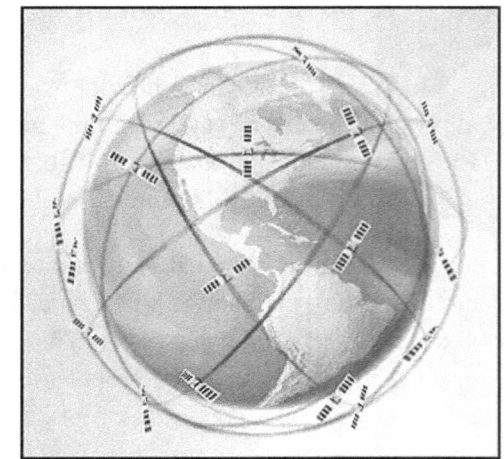

Figure 6-14: Globalstar

Globalstar service is beginning to be gradually introduced in each location once support systems have been completed and quality service is assured. There are 41 countries scheduled for initial service.

Users can place or receive calls using handheld, vehicle-mounted, or fixed terrestrial terminals. Globalstar's transmission data rate of 9600 bps can support numerous data services. Position location services can determine a user's location within 300 meters.

Globalstar coverage is limited to regions that lie within the same satellite footprint as a gateway station, and Globalstar associates have built far fewer gateway stations than originally planned. Users need to be in the footprint with a gateway station to complete a call. For example, service in Alaska will not be provided further north than Kodiak, as no gateway is near enough - although putting a gateway in Iceland provides service at slightly higher latitudes for North Atlantic shipping.

The first 4 satellites were launched in Feb 1998. The constellation was completed with 48 satellites in 1999. The satellites are in 764 nm circular orbits with 52 degrees of inclination. There are six active and one spare satellite in 8 planes. The satellite design life is 7.5 years. The satellites cover the globe from 70 degrees North latitude to 70 degrees South.

Orbital Communications (ORBCOMM)

ORBCOMM is the world's first commercial provider of global low-earth orbit satellite data and messaging services. The ORBCOMM system enables businesses to track remote and mobile assets such as trailers, railcars and heavy equipment; monitor remote utility meters and oil and gas storage tanks, wells and pipelines; and stay in touch with remote workers anywhere on the globe.

Each satellite is equipped with a VHF and UHF communication payload capable of operation in the 137.0-150.05 MHz and the 400.075-400.125 MHz bands. The ORBCOMM system is capable of sending and receiving two-way alphanumeric packets, similar to two-way paging or e-mail. Data rates of 2400 bps subscriber uplink and 4800 bps subscriber downlink are currently available.

The ORBCOMM constellation has 29 satellites in circular orbits 800 km above the Earth. There is 1 spacecraft in near polar orbit inclination of 108 degrees. The others are in orbits with 45 degree inclination. Unlike some of the other commercial LEO systems, ORBCOMM provides true worldwide coverage to include the polar regions. Pegasus is the launch vehicle used with seven to eight satellites being launched at one time.

FOREIGN SATCOM

There are far too many SATCOM systems to discuss in this text. Below is a list of some of the other SATCOM systems (other than Russia) and the country that owns them. The satellite may support countries other than the owning country.

COUNTRY	SATELLITE SYSTEM	COUNTRY	SATELLITE SYSTEM
Consortium	ARABSAT	Japan	JCSAT
Argentina	NAHUEL	Luxembourg	ASTRA
Australia	AUSSAT	Malaysia	MEASAT
Brazil	BRASILSAT	Mexico	SOLIDARIDAD
Canada	ANIK	Norway	THOR
Egypt	NILESAT	PRC	ASIASAT
Europe	EUTELSAT	Philippines	AGINA
France	TELECOM	Singapore	ST
Germany	DFS	South Korea	KOREASAT
India	INSAT	Sweden	SIRIUS
Indonesia	PALAPA	Thailand	THAISAT
Israel	AMOS	Turkey	TURKSAT
Italy	ITALSAT	UK	SKYNET

Table 6-2: Foreign SATCOM Systems

FUTURE SATCOM SYSTEMS

The DoD's vision for future SATCOM systems foresees advanced telecommunications services using Unmanned Aerial Systems, High Altitude Long Endurance (HALE) systems, multi-band satellite constellations, superior antenna and ground terminal technologies extending into a worldwide terrestrial fiber-optic network in a clean, seamless manner. The future systems supporting the warfighter must come on line without any degradation or gap in the quantity or quality of required communications. Commercial SATCOM systems will be heavily used when possible to take advantage of the rapid developments in technology.

The nature of future satellite communications systems will depend on the demands of the marketplace (direct home distribution of entertainment, data transfers between businesses, telephone traffic, cellular telephone traffic, etc.); the costs of manufacturing, launching, and operating various satellite configurations; and the costs and capabilities of competing systems - especially fiber optic cables, which can carry a huge number of telephone conversations or television channels. In any case, however, several approaches are now being tested or discussed by satellite system designers.

SATCOM CAPABILITIES

As discussed previously there are many things SATCOM can provide. This section will discuss general capabilities, the capabilities of the various frequency ranges, and the capabilities of the types of SATCOM.

General Capabilities

Satellites provide beyond-line-of-sight communications which are not significantly affected by man-made objects or natural terrain and can provide communications quickly to mobile units. In some situations, neighboring countries may not allow the positioning of ground relays on their territory. For example, during Operation DESERT STORM communications were required between headquarters in Saudi Arabia and Turkey. A ground based line-of-sight system would have required many terrestrial relays to be placed in Jordan and Syria or in Iran. Even if those countries would have allowed US ground relay stations on their soil, security and logistics for such remote sites would have been very difficult.

SATCOM can provide long range communications to remote areas without the need to establish numerous intermediate ground relay links thus communications is available to military forces worldwide, 24 hours a day. Communications satellites are highly reliable and typically operate in space for many years. SATCOM can provide a high volume of signal traffic and high quality circuits. SATCOM terminals are generally smaller and more deployable than the components of comparable non-SATCOM long distance communication systems.

The cost of transmitting information between two users via satellites is essentially the same despite the distance. A signal can be relayed across the country or across the ocean by satellite as cheaply as across the street by satellite. Satellites can be used as broadcast transmitters. Information can be relayed and received over a wide area. Satellites can deliver large amounts of information within a given amount of time

Most satellites can be moved or its antennas aimed to provide better coverage of specific areas. Satellite communication uplinks have a relatively low probability of being intercepted. Directional ground antennas can make transmissions to satellites very hard for an enemy to detect. Many SATCOM satellites are in a geosynchronous orbit and are therefore more survivable against anti-satellite weapons because they are so far away. Lastly, the unique capabilities of satellites are rapidly giving rise to new communications concepts bringing more information more quickly to the warfighter.

Capabilities by Type of SATCOM

Single channel SATCOM supports a variety of missions because of the flexibility and mobility provided. For the warfighter on the ground who is already burdened with large amounts of life support gear, the use of small, portable, single channel SATCOM terminal is very beneficial. Single channel SATCOM is extremely useful in deployment, entry, and short duration operations.

Access to commercial SATCOM is usually less difficult than military SATCOM access, barring any host nation approval problems. The Army cannot satisfy its myriad communications requirements without commercial SATCOM.

The advantages of multi-channel satellite communications are extensive. Although submarine cables, fiber optics and microwave radio can effectively compete with satellites for geographically fixed wide-band service, the satellite is unchallenged in the provision of wide-band transmissions to mobile terminals. Satellite communication systems can connect with other military communications systems such as Mobile Subscriber Equipment (MSE), other multi-channel systems, FM radio and radio teletype thus extending the range of users. The inherent flexibility that a satellite communications system provides is essential to the conduct of military operations both nationally and globally.

Commercial SATCOM is readily available and offers better interoperability among multi-national forces.

SATCOM LIMITATIONS

There is really no viable alternative to satellite communications for military applications. However, SATCOM has limitations that must be considered.

Limitations in General

From a terrestrial perspective, SATCOM is considered a beyond-line-of-sight system. There must, however, be line-of-sight between the terrestrial terminal and the satellite. Terrain can still affect line-of-sight especially when the angle between the terminal and the satellite it must communicate with is very low. Spot beam downlinks from some communications satellites limit where the signals can be received and downlinks are more susceptible to interception than direct line-of-sight ground systems or cable systems. Broadcast communications can be received by any terminal capable of operating in the frequency band with the correct cryptographic key. Communications on-the-move is possible but still limited.

GEO SATCOM is not easily moved to provide more coverage to the area of operation. Any moves beyond the programmed moves budgeted in the satellites' design life are time consuming and use fuel that was meant to keep the satellite in orbit thus shortening its life span. GEO SATCOM does not provide communications coverage of the poles. LEO SATCOM doesn't have the polar coverage problem; however, because of the lower orbit, the life span of the satellites is shorter and it is more susceptible to anti-satellite weapons.

Limitations by Frequency

Access to UHF SATCOM channels is tightly controlled. The capacity is relatively low; however, it is adequate to support single channel voice circuits. The long-awaited implementation of Demand Assigned Multiple Access (DAMA) will increase the capacity of the UHF band and provide more access to the channels. UHF signals can be easily detected and jammed by enemy forces and they are disrupted by natural scintillation. UHF has a relatively low capability to resist jamming of the satellites. Current technology UHF signals are easily detected and intercepted.

The cost of the SHF terminals and the satellites is higher than that for UHF, but many of the systems have already been acquired; therefore, future costs should be moderate. Mobility is decreased with DSCS III because of the size of the user terminals and antennas, and the power needed to operate them. Army terminals are deployable but require more than one truck per terminal.

EHF satellite and terminal costs are higher than the other frequency spectrums because it is relatively new technology. A new family of user terminals is needed which will increase the overall cost of the system. The designed capacity of the Block I Milstar satellites is relatively low. EHF is extremely degraded in heavy foliage and rain (rain fade).

Limitations by Type of SATCOM:

Single channel SATCOM has a number of limitations the foremost of which is limited capacity. Current ground terminal limitations include a lack of communications on-the-move capability and difficulty in voice recognition over single channel SATCOM systems. Many single channel terminals use UHF frequencies. Milstar single channel terminals will use the EHF band.

Commercial SATCOM has high costs associated with access and it is often difficult to get host nation approval for use. Commercial SATCOM is not jam resistant or nearly as survivable.

PLANNING CONSIDERATIONS

What are the communications requirements? Can they be satisfied with terrestrial systems?

The warfighter must understand who needs the communications, under what conditions, and in what kind of terrain. SATCOM can fill many of the warfighter's requirements for communications to forward units in remote areas, command and control of mobile units, and to units deployed in areas that do not have telecommunications systems available. However, the type of SATCOM used may be different for each situation. Commercial SATCOM can be used when protection of networks is not a priority or the network is designed to provide administration, logistics, or humanitarian/disaster relief communication support. Single channel UHF can be used in dense foliage where single channel EHF cannot. SHF or EHF multi-channel can support requirements for large volumes of data. UHF can support broadcasted information requirements over the area of operation.

The CJCS "owns" all SATCOM assets and apportions them geographically to each Combatant Commander who then owns that portion of SATCOM within their theater. The Combatant Commander then apportions out those resources to the J/CTF and/or component commands depending on the warfighting scenario and priorities. The Army Satellite Communications Architecture Book and CJCSI 6250.01 give all the necessary information about how to request communications satellite access.

Listed below are some questions for leaders and communications planners. The following sections about capabilities and limitations can be used to determine if SATCOM meets the communications requirements, does it need to be MILSATCOM or will commercial SATCOM suffice, and which frequency range meets the requirements. Requirements should be pushed to J6 for resolution

- What is the mission?

- What types of circuits are required?

- How critical are each of those circuits?

- What information will be passed over the circuit?

- How much capacity is needed?

- Are there validated requirements to support the mission?

- What allied systems must be supported or interfaced with?

- What kinds of ground terminals are available?

- Are there host nation agreement problems?

- What communications mediums are available during each phase of the operation?

- What can the telecommunications network in the operational area provide?

- What MILSATCOM assets are available and will they fulfill the requirements?

- Are leased satellite communications required and/or available?

- Are there international agreements that prohibit their use?

- Does the military have "first rights of refusal" agreements with commercial satellite providers for service if needed?

- Do marginal or residual capabilities on spare satellites need to be activated? Are there test satellite assets available?

- What kind of terrain, climate, vegetation, and/or buildings may be present that will interfere with terrestrial or SATCOM equipment?

- What are the jamming or interception capabilities of threat forces?

- What are the SATCOM capabilities of the threat forces? How will commercial SATCOM support the threat?

Chapter 7
Position, Velocity, Navigation, and Timing

OVERVIEW

Satellite navigation systems, such as the US Navigation Satellite Timing and Ranging Global Positioning System Global Positioning System (NAVSTAR GPS) came of age with the Gulf War of 1991. GPS, a space-based radio navigation system, supported coalition forces in targeting, navigation, reconnaissance, refueling, air- and sea-launched cruise missiles and providing the Army logistics forces with accurate navigation across the trackless desert to keep up with moving ground forces. Today, GPS is used in a variety of applications, both military and civilian, and the uses are continually expanding.

A space-based radio navigation system uses radio transmissions for location determination. Unlike previous navigation systems using ground based transmitters, satellite based transmitters are used to cover earth with higher accuracy than that of land based systems. The satellites transmit timing information, satellite location, and health information. The receiver contains a specialized computer that calculates the location based on the satellite signals. Nothing except a receiver and the satellite signals are needed to use the system, so it is immediately available to users as they deploy into any theater of operation. The user does not transmit anything to the satellite, and the satellite does not know the user is there. There is no limit to the number of users that can be using the system at any one time.

In 1960, Transit, the first navigation satellite, was developed by the Applied Physics Laboratory of Johns Hopkins University for updating the inertial navigation systems of the US Navy's Polaris submarines. Transit Systems provided navigational support to the US Navy and commercial users and it was intended to provide 2-dimensional navigation data. At the same time, the Air Force conducted concept studies for a 3-dimensional navigation system called 621B. There was concern that these systems were duplicating capabilities. The Air Force was designated as the executive agent to consolidate the Timation and 621B concepts into a comprehensive system which could meet the requirements of all the services. The Transit System was deactivated in December 1996 after the (NAVSTAR GPS) constellation was declared fully operational on 27 April, 1995.

NAVSTAR GLOBAL POSITIONING SYSTEM (GPS)

The NAVSTAR Global Positioning System is a space-based, all weather, continuous operation, radio navigation system. The system provides military, civil, and commercial users highly accurate worldwide three-dimensional, common-grid, position/location data, as well as velocity and precision time to accuracies that have not been easily attainable before.

Management and oversight of dual-use aspects of GPS (military and civil use) is provided by the National Executive Committee for Space-based Positioning, Navigation, and Timing jointly chaired by the DoD and Transportation. They manage the GPS and US Government augmentations. Other departments and agencies participate as appropriate.

The DoD is required to: acquire, operate, and maintain the basic GPS; maintain a Standard Positioning Service that will be available on a continuous, worldwide basis; and maintain a Precise

Positioning Service for use by the US military and other authorized users. From the program's inception in the 1970s, the DoD has been dedicated to successful management of the GPS as a dual-use national information resource. DoD's stewardship of GPS has been instrumental in the growth of a new global industry. The Department works in this management structure to maintain the delicate balance between global security and economic interests in the operation of GPS.

The Department of Transportation is responsible for serving as the lead agency within the US Government for all Federal civil GPS matters. It is also responsible to develop and implement US Government augmentations to the basic GPS for transportation applications.

Segments

GPS does not refer only to the NAVSTAR satellites that orbit the Earth. It consists of three distinct segments: the Space Segment, the Control Segment, and the User Segment. All three have a role to play in providing users with accurate position, velocity, and timing data.

Space Segment

The Space Segment is designed as a full constellation of 24 satellites in 6 circular orbital planes with an inclination of 55 degrees. Each plane contains 4 satellites at 10,900 nautical miles altitude with a period of about 11 hours and 58 minutes. These orbital planes maximize coverage of the earth with slight degradation at the poles. The semi-synchronous orbit creates a constantly changing user-to-satellite relation.

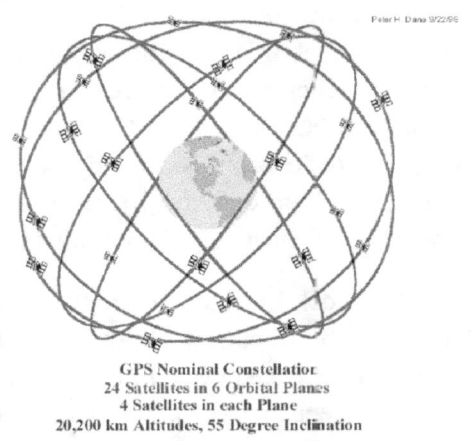

GPS Nominal Constellation
24 Satellites in 6 Orbital Planes
4 Satellites in each Plane
20,200 km Altitudes, 55 Degree Inclination

Figure 7-1: GPS Constellation

The way the satellites are phased within the constellation allows for loss of several satellites with minimal impact on users. This is often called "graceful degradation." The constellation capability will "degrade gracefully," meaning users should only lose a part of their navigation capability as they lose a satellite.

The design life of the current GPS satellites is seven years. Some have continued to function for more than 10 years. Replacement satellites are launched, as necessary, to replace ones that begin to develop problems or with the design of major modifications to the satellites. The constellation has had as many as 32 operational satellites at one time.

The GPS signals are transmitted continuously by all the GPS satellites. With the fully operational system, users anywhere in the world can receive signals from at least five to eight satellites at all times. Usually, six GPS satellites will be in view. If one satellite should fail to provide accurate data, there are normally sufficient satellites to give total coverage. As a result, availability of the system is estimated at more than 99% of the time.

Figure 7-2: GPS IIR

GPS Block IIR satellites began replacing the older GPS Block II/IIA satellites in 1997. The GPS Block IIR satellites boast dramatic improvement over the previous blocks with reprogrammable satellite processors to enable fixes and upgrades in flight. Block IIR satellites are designed to provide at least 14 days of operation without contact from the Master Control Station and up to 180 days of operation when operating in the autonomous navigation (AUTONAV) mode. Full accuracy is maintained using a technique of ranging and communication between the Block IIR satellites. The cross- link ranging is used to estimate and update the parameters in the navigation message of each Block IIR satellite without contact from the Control Segment. The design life of the Block IIR satellite is 7.8 years; each contains three atomic clocks: two rubidium (Rb) and one cesium (Cs); and have the Selective Availability (SA) and Anti-Spoofing (A-S) capabilities. These will be discussed later. Eight of the remaining Block IIR satellites will also be specially modified as GPS IIR-M satellites, the "M" standing for modernized. The new features will not be taken advantage of until receivers are designed and proliferated and there are enough of these satellites in the constellation to take advantage of these new capabilities.

The first GPS IIR-M was successfully launched on 25 September 2005 from Cape Canaveral, FL. This modernized series offers a variety of enhanced features for GPS users, such as a modernized antenna panel that provides increased signal power to receivers on the ground, enhanced encryption and anti-jamming capabilities for the military, a second civil signal that will provide civil users with an open access signal on a different frequency (L2C) and a new military signal (M-Code) on both the L1 and L2 channels. L2C enables the development of lower-cost, dual-frequency civil GPS receivers that allow for correction of ionospheric time delay errors. In the near future, enhancements such as dataless and pilot channels for improved performance and an improved navigation message with more precise clock and ephemeris information will be available.

GPS Block IIF satellites are the next generation of GPS space vehicles. The GPS Block IIF satellites, being built by Boeing, will provide all the capabilities of the previous blocks with additional benefits to include an extended design life of 12 years, faster processors with more memory, and a new civil signal on a third frequency (L5). L5, which will be broadcast beginning with the first IIF satellite, will lie in the "Aeronautical Radionavigation Services" band and can be used for safety-of-life aviation. It will be compatible with Galileo, GLONASS, and Quasi-Zenith Satellite System QZSS (under development by Japan), with the goal to be interoperable as well. L5 will transmit at a higher power than current civil GPS signals, and have a wider bandwidth. Its lower frequency may also enhance reception for indoor users. The first GPS Block IIF satellite is scheduled to launch in 2009.

GPS Block III will be the next block of GPS satellites. GPS IIIA will transmit a new civilian signal (L1C), which is designed to be highly interoperable with the European Galileo satellite navigation system signal, if and when the ESA establishes that constellation.

Control Segment

The Control Segment works as a system to ensure the overall health and accuracy of the GPS constellation. It is composed of Monitor Stations (MS), the Master Control Station (MCS), and Ground Antennas (GA). The Monitor Stations measure the same GPS signals as users. The Monitor Stations are GPS receivers at fixed sites that take accurate measurements from all GPS satellites in view and send the data along with satellite clock data to the MCS for processing and error detection. The MCS

146

uses these measurements to calculate errors in the constellation and generate more accurate navigation information for each satellite. Operators in the MCS calculate each satellite's status, ephemeris and clock data which is then sent to transmitting antennas located at the Monitor Stations (except Hawaii) where the data is uploaded to each satellite for inclusion in the navigation message transmitted by the satellites. This is done to maintain the desired system accuracy. The GA's are the interface between the MCS and the Space Segment. They are used to obtain telemetry from the satellites and to transmit commands and navigation information to the satellites. The GA's are located worldwide to ensure we can contact any satellite with minimal delay.

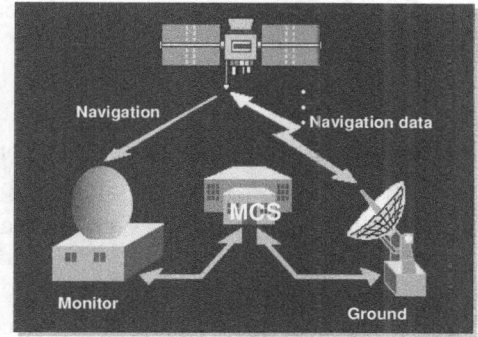

Figure 7-3: Control Segment Interactions

Continuous constellation monitoring is vital for keeping GPS accurate. The constant loop of collecting measurements, processing information, and transmitting to the satellites is all accomplished by the Control Segment.

Figure7-4: GPS Control Segment Locations

The Control Segment is operated by the 50th Space Wing of Air Force Space Command. The GPS MCS and a Monitor Station are located at Schriever Air Force Base, Colorado Springs, Colorado. Four other Monitor Stations are located in Hawaii, Ascension Island, Diego Garcia and Kwajalein Atoll. The three GA's are located at Ascension Island, Diego Garcia and Kwajalein Atoll. As of 2005, the National Geospatial-Intelligence Agency (NGA) has also tied several of its ground sites into this monitoring/update network for GPS.

Sometimes satellites are taken "off the air" due to maintenance of the atomic clocks, orbital maneuvers, or address problems on the satellite. The MCS tries to ensure there is never more than one satellite "off the air." Most maintenance is a scheduled event and a daily notice is published by means of a Notice Advisory to NAVSTAR Users (NANU). It is possible to coordinate maintenance with the MCS so that taking a satellite temporarily out of service does not have an adverse effect on critical may operations.

The US Coast Guard Navigation Center Navigation Center (NAVCEN) serves as services for safe, secure, and efficient maritime transportation by delivering: enhanced situational awareness through tracking vessel movements, quality positioning, navigation and timing signals, accurate and timely maritime information services, and operational and technical oversight of eNavigation systems.

NAVCEN operates the Coast Guard Maritime Differential GPS (DGPS) Service and the developing Nationwide DGPS Service (NDGPS), consisting of two control centers and 86 remote broadcast sites. The Service broadcasts correction signals on marine radiobeacon frequencies to improve the accuracy and integrity to GPS-derived positions. The Coast Guard DGPS Service provides 10-meter accuracy in all established coverage areas.

User Segment

The User Segment is composed of anyone who uses the navigation signal from space. Since the signals are broadcast worldwide, the users are worldwide. There are two general categories of users: military (authorized) and civilian (non-authorized). This doesn't mean that civilians can't use the GPS signal; in fact they are the largest user segment. The growing civilian community's reliance on GPS is becoming a major issue for constellation operations, hence the development of new civilian signals on the latest satellites.

The development and acquisition of GPS receivers for the military users is managed by the GPS Joint Program Office (GPS JPO) located at Los Angeles Air Force Base, California. The GPS JPO is staffed by personnel from the US Air Force, Army, Navy, Marine Corps and Coast Guard along with representatives from the US National Geospatial-Intelligence Agency (NGA), Australia, and many NATO countries. The GPS JPO also provides information to manufacturers of civilian GPS receivers and processors.

Military receivers include several types of portable handsets and aircraft units. The Precision Lightweight GPS Receiver (PLGR) was the standard Precision Positoining Service (PPS) hand-held unit. More than 150,000 were delivered to DoD. The Defense Advanced GPS Receiver is the current, standard hand held receiver. Some of military receivers that support Military operations include:

- Precision Lightweight GPS Receiver (PLGR)

- GPS 3A Receiver

- Defense Advanced GPS Receiver (DAGR)

- Miniaturized Airborne GPS Receiver (MAGR)

- GPS Receiver Applications Module (GRAM)

- Cargo Utility GPS Receiver (CUGR)

- Special Operations Lightweight GPS Receiver (SOLGR)

- Standalone Air GPS Receiver (SAGR)

The number of channels in a receiver determines how many satellites it can receive signals from simultaneously. When not moving, the number of channels in a receiver is not a major factor in determining its position accuracy. When stationary, a 1-channel GPS receiver can be just as accurate as a 5-channel receiver can. More channels do allow a receiver to respond and update position solutions faster and multiple channels do allow a receiver to remain locked on the satellites in view. A multiple channel set does provide better performance when moving, especially when moving very slowly or very fast because a single channel set does not update its data quickly enough to detect small changes in position. The result can lead to inaccurate direction of movement while moving dismounted in a field environment. A multiple channel receiver detects small changes in position with greater distinction, thus it is able to report a more accurate and consistent direction of movement. A multiple channel set will also reacquire satellites faster following a brief interruption in the signal.

HOW GPS WORKS

Missions

GPS missions are:

- Navigation - primary mission

- Time Transfer - subset of navigation mission

- Nuclear Detonation Detection - secondary mission

Navigation:

GPS is based on the concept of trilateration (similar to triangulation, but in three dimensions instead of two and no angles are measured) from known points similar to the technique of "resection" used with a map and compass except that it is done with radio signals transmitted by satellites. The user's GPS receiver must determine precisely when a signal is sent from selected GPS satellites and the time it is received. The receiver measures the time required for the signal to travel from the satellite to the receiver, by knowing the time that the signal left the satellite, and observing the time it receives the signal, based on the receiver's clock.

The navigation mission is accomplished by broadcasting time-synchronized signals from the constellation of GPS satellites. A GPS receiver has to acquire and track signals from GPS satellites, achieve carrier and code tracking, collect data from the NAV message included in the signals, and then make pseudorange and relative velocity measurements. From these the receiver can calculate the GPS time, its position and velocity. For a receiver on the ground to determine its 3-dimensional position it must calculate four unknowns: latitude, longitude, altitude and time. For this reason the ground receiver must receive signals from four satellites. If the receiver had a perfect clock, exactly in sync with those on the satellites, three measurements, from three satellites, would be sufficient to determine position in 3 dimensions. Unfortunately, we can't get a perfect clock that will fit (financially or physically) in a $300 (or even $3000) receiver, so a fourth satellite is needed to resolve the receiver clock error. Each measurement ("pseudorange") gives the size of a sphere centered on the corresponding satellite. The four satellites generate a solution of the intersection of these four spheres. Due to the receiver clock error, the four spheres will not intersect at a single point, but the receiver will adjust its clock until they do, providing very accurate time, as well as position. To calculate a 2-dimensional position (no altitude) requires three satellites to be in view of the receiver. With a full constellation the receiver can view six satellites from most locations and can view a maximum of twelve satellites from some locations.

To get the best possible accuracy, a GPS receiver will select the satellites that offer the best geometry. This is the same approach that soldiers use in selecting points to sight on when using the technique of resection with a map and compass to determine a location. A more accurate answer is obtained by sighting on two or more points that are far apart. This is also true with GPS. Satellites that appear farther apart in the sky provide a more accurate position solution than satellites that are close together. Since the ephemeris (computed position) of each satellite is known by the GPS receiver from data obtained from each satellite's navigation message, it is possible to calculate which combination of GPS satellites provide the best geometry at a given time.

The accuracy of GPS receivers is stated in statistical terms. It is important to have some understanding of these terms so that the data, particularly the accuracy of positions, is not misinterpreted. Many GPS receivers can display 10 digits in MGRS grid coordinates which equals to 1-meter resolution. This does not mean that the receiver has 1-meter accuracy. Manufacturers and the military often use different techniques to express accuracy. When comparing performance, the comparison must be made under the same operating conditions and expressed in the same terms. Precise conversion between these different techniques requires a rigorous statistical solution; however, it is possible to give some approximate equivalents. A normal distribution is assumed.

Linear Error Probable (LEP) is defined as the distance from a point on a line within which 50% of the measurements will occur. Usually LEP is used to express the vertical (altitude) error.

Circular Error Probable (CEP) is defined as the radius of a circle containing 50% of the individual measurements. A receiver with an accuracy of 100 meters CEP means that 50% of the time the solution will be correct within a radius of 100 meters and 50% of the time the error will be greater than 100 meters. CEP usually refers to accuracy in the horizontal plane only without regard to vertical (altitude) accuracy.

Spherical Error Probable (SEP) is defined as the radius of a sphere within which there is a 50% probability of locating a point or being located. SEP includes both horizontal and vertical error.

2-Dimensional Distance Root Mean Squared (2dRMS), as defined in STANAG 4278, is the radius of a circle that contains 63% of all measurements. For example, 100 meter (2dRMS) means that 63% of all solutions will be within a circle with a radius of 100 meters. There are, however, some documents that base 2dRMS on a 95% probability level. An accuracy capability of 100 meters (2dRMS @ 95%) is better than 100 meters (2dRMS @ 63%).

Time Transfer

A sub-mission of the navigation mission that is critical to communications and financial transactions is the time transfer mission. Time transfer allows users to calculate and synchronize time with great accuracy simultaneously around the world. Each GPS satellite has atomic clocks on board to maintain accurate time. The accuracy of these atomic clocks is sent to the MCS and corrections are sent to the satellites whenever necessary to keep the system within specification. Atomic clocks are not nuclear powered; they get their name because they use the very stable oscillations of certain elements, in the case of GPS, rubidium or cesium, to measure the passage of time. The accuracy is very high (+/- 1 second every 360,000 years) but it is not perfect. Since the receiver must adjust its clock to be precisely in sync with GPS time, a GPS receiver can be used as a precise time reference. Some receivers provide a 1 pulse per second output for this purpose.

The atomic clocks are synchronized so that the GPS navigation signal is sent from each GPS satellite at precisely the same time. The specification for accuracy of the GPS atomic clocks provides for time transfer to within 100 nanoseconds of the United States Naval Observatory's (USNO) Universal Time Coordinated (UTC). Although the specification states that the time transfer should be within 100ns, the actual performance of time transfer has been averaging about 10ns. Data in the navigation message tells how far GPS time is off from USNO time so the user can calculate the UTC time to within the 100ns mentioned above.

One of the primary uses of the time transfer is for synchronizing digital communications. Examples would include: SINCGARS and HAVE QUICK. SINCGARS is a survivable radio used by the Army and the Marine Corps. HAVE QUICK is a radio that employs a frequency hopping scheme for anti jam capability. The "hop" must occur at the same time as all the other HAVE QUICK radios. This requires accurate time synchronization. GPS can provide the correct time through a GPS user set. The Milstar constellation also uses GPS time synchronization to ensure proper relay of its digitized communications.

All GPS position/navigation receivers calculate the precise time in order to determine location, however, most only display the time to no more than 1/100th of a second. There are special receivers that calculate the time much more precisely.

Nuclear Detonation Detection

The secondary payloads onboard the GPS satellites include a variety of sensors used to detect nuclear detonations (NUDETs). The large number of satellites ensures detection of any event worldwide. A nuclear detonation can be located to within 1.5 km. The system will sense NUDETs and pass the information to the Air Force Technical Applications Center (AFTAC) downlink locations at Buckley AFB, CO (Denver) and Schriever AFB, CO (near Colorado Springs).

Signals, Codes, Services

In general, the responsiveness and accuracy of GPS receivers is determined by the electronics (hardware) and programs (software) stored in the set. Innovative approaches by manufacturers are common.

There are different signals, codes and services available over GPS. The signals, codes, and services are mutually exclusive entities, yet entirely interrelated. If a user has one, he does not necessarily have the other (e.g., if the user has P-code it doesn't mean he has precise positioning service) but needs to know one to explain the other. There are three GPS signals and two codes broadcast at all times. From this there are two basic types of service available.

Signals

The GPS currently has signals broadcast over four L-Band frequencies (L1, L2, L3, and L5). L1 and L2 provide navigation information. L3 is only used to transmit NUDET information and has no impact on navigation users. L5 is the new additional signal to assist commercial navigation accuracy but is not universally available as yet. The primary advantage of getting both L1 and L2 frequencies is that can reduce propagation error through the ionosphere. This is done by dynamically measuring differences in refraction caused by the different frequencies. If the receiver does not get both frequencies, it must rely on a less accurate ionosphere model variable factor included in the navigation message.

Codes

The two codes are the Coarse Acquisition Code (C/A-code) and the Precision Code (P-code). As of 7 June 1996, all GPS satellites broadcast both the C/A-code and P-code on L1 and only the P-code on L2.

- The C/A-code consists of a 1023 bit code with a clock rate of 1.023 MHz, hence, it is easy to synch up and acquire. The C/A-code is used to provide Standard Positioning Service. The C/A-code is relatively short and repeats itself every millisecond. A unique C/A-code is assigned to each GPS satellite so that receivers can distinguish among them. The C/A-code is available to all users.

- The P-code is a 267 day long code sequence. Each satellite has a unique seven-day section of the code transmitted with a 10.23 MHz bit rate. The C/A-code is used by P-code users to assist the receiver in reducing the time to acquire the longer P-code. The P-code receivers will generally lock up on C/A-code and read out a handover word to synch up on the P-code. The P-code is more jam-resistant due to the increased bandwidth used for the spread-spectrum scheme. The C/A-code bandwidth is 2.046 MHz, while the P-code bandwidth is 20.46 MHz. The P-code is only available to users authorized by the DoD.

- The P-code is protected against spoofing (i.e., the deliberate transmissions of incorrect GPS information) by encryption of the P-code. When this is done, the P-code is called Y-code. Anti-Spoofing (AS) encrypts the P-code into the Y-code to prevent an enemy from transmitting signals which could mimic a GPS satellite. The same cryptological key used to remove the effects of Selective Availability (SA) is used to decrypt the Y-code into the P-code. The Y-code is also only available to authorized users. SA was implemented on the C/A-code and could also be implemented on the P-code. SA and AS can be implemented independently. Both are discussed in more detail later below.

Services

The two types of service available are Standard Positioning Service (SPS) and Precise Positioning Service (PPS). SA is associated with the services available.

- SPS is available to all users around the world. Users do not require any special codes other than standard C/A-code and access to the SPS does not require approval by the DoD. The accuracy of GPS may be degraded for SPS users, but it is normally within 100m of the horizontal position (2-Dimensional root mean square) 95% of the time. The time transfer function will be within 340ns 95% of the time. In a time of crisis this accuracy could be degraded more significantly to prevent an enemy from using GPS to its advantage. The amount of error, called SA, induced would be based on DoD policy and decisions. The President will make decisions to change accuracy levels.

- PPS is a highly accurate position, velocity, and timing service that is only made available to authorized users. Authorized users are those who are designated by the Assistant SECDEF for C3I and who have the crypto keys required to remove SA errors. PPS provides the GPS receiver access to the most accurate signals from the satellites. The 3-D positioning accuracy is guaranteed to 16m Spherical Error Probable at least 50% of the time. This translates to 3-D position within 30m, 95% of the time. The timing accuracy is within 100ns 50% of the time within 197ns 95% of the time. The velocity accuracy is 0.1 meters/second RMS. PPS users must have the keys and equipment to decrypt Y-code and remove SA error. PPS users usually have a dual frequency capability based on P(Y)-Code on both L1 and L2.

The key difference between the SPS users and the PPS users is that PPS users have the crypto keys to remove SA errors and decrypt the Y-code. Also of note, most PPS users have a dual L1 and L2 frequency receiver, which allows for better atmospheric error correction and in turn better accuracy. A comparison of SPS and PPS is shown in the table below.

GPS ACCURACY		
	Standard Positioning Service (SPS)	Precise Positioning Service (PPS)*
Position	76 m SEP 40 m CEP 100 m 2drms @ 95%	16 m SEP 9 m CEP
Velocity	0.5 m/sec	0.1 m/sec
Time	1 millisecond	100 nanoseconds
* PPS is only available to US and allied military, US Government and selected civil users specifically approved by the US Government.		

Table 7-1: GPS Accuracy

Selective Availability Management

The main purpose of Selective Availability Management was to prevent unauthorized users from using the capability of GPS against the US while ensuring that authorized users can use it. The accuracy of GPS would be degraded for SPS users if SA is used, but by Presidential decision, SA has been set to zero since 1 May 2000 so it is currently not a factor. Normally GPS accuracy should be at least within 100m of the horizontal position (2-Dimensional root mean square) 95% of the time. The Standard Positioning Service (SPS) available to civilian users most likely will give 20 meter horizontal accuracy with SA off. The vertical accuracy is about 1.5 times worse than horizontal, due to satellite geometry. (Satellites are more likely to be near the horizon, than directly overhead.) There are 2 components to Selective Availability Management: Selective Availability and Anti-Spoofing.

Selective Availability (SA)

Selective Availability injected known errors into the satellite signal. The injected errors resembled what was required for accurate trilateration. PPS users with current cryptographic keys loaded in their receivers decrypt corrections to the SA errors to regain high precision accuracy. SPS can provide very good accuracy for most applications.

Users authorized by the DoD to use GPS are provided with a cryptological key which can be loaded into GPS receivers. The key removed the effects of SA and allows the receiver to calculate the best solution possible. During a time of crisis, the error caused by SA could be increased to about 2,000 meters, thus making GPS much less worthwhile. SA can be applied to both the C/A-code and the P-code.

It is US policy that we will continue to provide the GPS Standard Positioning Service for peaceful civil, commercial and scientific use on a continuous, worldwide basis, free of direct user fees. We will cooperate with other governments and international organizations to ensure an appropriate balance between the requirements of international civil, commercial and scientific users and international security interests. Lastly, we will advocate the acceptance of GPS and US Government augmentations as standards for international use.

Anti-Spoofing (AS).

AS is different from SA in that it is implemented to circumvent erroneous impersonation of GPS signals by hostile forces. AS replaces the normal P-code with an encrypted version called the Y-code. The Y-code can only be decrypted by authorized users. Only the P-code is affected. C/A-code is still subject to spoofing.

Sources of Navigation Error

As with any measuring system, GPS cannot provide absolute precision because errors are introduced from a number of different sources. These sources of error are summarized below. The next table shows the typical extent of the error.

The orbits of GPS satellites have been selected to optimize stability, longevity and coverage. The orbits are very stable; therefore it is possible to calculate each satellite's ephemeris with high precision. Even so, slight variations are introduced due to the uneven density of the Earth, magnetic fields in space, fluctuations in solar radiation and other factors external to the satellites. In addition the ephemeris prediction model is not absolutely precise. The monitor stations track each satellite and the GPS control segment updates the ephemeris frequently (usually every four hours) to remove as much error as possible.

GPS receivers use advanced electronic circuitry to receive, decode and process the data sent by the satellites. The receivers are not, however, perfect so a small amount of error is introduced. As stated before, if the receiver had a perfect clock, exactly in sync with those on the satellites, three measurements, from three satellites, would be sufficient to determine position in 3 dimensions.

A certain amount of error is caused by reception of multipath signals. Multipath results from signals being reflected off of objects in the vicinity of the receiver (e.g. buildings in an urban environment). Most receivers employ techniques to minimize the impact of multipath signals.

The ionosphere is made up of charged particles. The size and density of the ionosphere over a particular area is always changing due to sunlight, fluctuations in the Earth's magnetic field, solar radiation and other factors. Radio signals transmitted through the ionosphere are slowed in a manner related to the frequency of the signal and the state of the ionosphere at that instant. The lower the frequency of the signal, the more it is slowed by the ionosphere. Uncorrected ionospheric delay results in significant error in the position solution. PPS receivers receive the P-code on both the L1 and L2 frequencies. Since the same code is transmitted at the same time on two different frequencies it is possible to measure the relative difference in the time of arrival and calculate the effect of the ionosphere. SPS receivers cannot process the P-code; therefore an iononspheric model is used to estimate the ionospheric delay; however, this is less accurate than the PPS.

GPS Augmentation

Although GPS is, by itself, a very capable system, it is possible to augment the basic system so that the end result is even better. The idea behind GPS Augmentation is to obtain greater accuracies from GPS by using a second source to improve upon the GPS data. This section will give an idea of some of the current augmentation sources. It will also discuss ways to improve upon GPS without outside sources.

There are a variety of sources for GPS Augmentation. Some sources are Differential GPS, Wide Area Augmentation System, Local Area Augmentation System, and GPS Aiding.

Differential navigation (DGPS)

Differential GPS (DGPS) is a means to eliminate the effects of Selective Availability and correct for some GPS errors by using the errors observed at a known location to correct the readings of a roving receiver. The basic concept is that the reference station "knows" its precise position, and determines the difference between that known position and the position as determined by a GPS receiver. The reference stations may be a permanent service, or setup specifically for a project. It measures the distances to each satellite and calculates the errors associated with the satellite. The satellites are so high up, that any error measured by the differential station wall be essentially the same for any receiver in the local area. It then transmits error corrections to any properly equipped GPS differential receiver in the area. When applied at the roving receiver, these corrections greatly enhance its positional accuracy. The correction can be in terms of

Figure 7-5: East Coast DGPS

position, or more often in terms of the observed satellite-receiver distance (the pseudo-range). The corrections may be collected and applied at a later time, or they may be broadcast immediately to the roving receiver by mobile phone, radio or satellite communications.

To use differential GPS, a user must have a receiver able to use corrections transmitted from the Differential Station and GPS satellites. DGPS will eliminate the error introduced by Selective Availability, and errors caused by variations in the ionosphere, resulting in reported positions within about 10 meters (33 ft.) of the true position 95% of the time. Better receivers can get within 3 meters, or so. The one error correction from the differential station corrects for all the errors in the GPS signal: receiver clocks; satellite clocks; satellite position; ionospheric delays; atmospheric delays. The accuracy of DGPS, of the order of a few meters, generally degrades with increased distance from the nearest base station.

For marine use, the US and Canadian Coast Guards (and corresponding agencies in other countries) have established DGPS reference stations that broadcast the correction data over the existing 250 - 350 KHz marine radio beacons. This marine service is available free of charge in the US and Canada, but may only be available by subscription in some countries. The DGPS correction data can be used as far as 1500 km from the reference station depending on the DGPS setup -- if the DGPS is part of a larger monitoring network.

Differential GPS has potential for applications requiring high accuracy. Some examples are:

- Field artillery and mapping survey.

- Delivery of precision munitions.

- Marking obstacles and cleared paths to follow.

- All weather helicopter operations.

- Non-precision approach for aircraft, particularly into non-instrumented airfields.

- Autonomous operation when the control segment cannot update the satellite ephemeris.

- Instrumented training ranges.

The Coast Guard's maritime Differential Global Positioning Service achieved Full Operational Capability (FOC) on 15 March 1999. NAVCEN operates the Coast Guard Maritime Differential GPS Service, consisting of two control centers and over 50 remote broadcast sites. The Department of Transportation's Nationwide Differential GPS (NDGPS) expansion is underway. The NDGPS plan calls for the conversion of a number of US Air Force Ground Wave Emergency Network (GWEN) sites in their current location and relocation of the remaining sites into desired regions.

Wide Area Augmentation System

The Wide Area Augmentation System (WAAS) is an expansion to DGPS. Rather than transmitting corrections within a local area, the corrections will be transmitted across the entire US and potentially worldwide.

WAAS is a safety-critical navigation system that will provide a quality of positioning information never before available to the aviation community. Developed by the Federal Aviation Administration (FAA), this system uses a number of reference stations (WRS) scattered around the US. These correspond (somewhat) to the differential correction stations of the Coast Guard marine DGPS system, but do not transmit the correction signals themselves. They monitor the GPS signals, ionospheric conditions, and the WAAS correction signal, and transmit the data to the WAAS master stations (WMS). The WAAS master stations take the data, validate past correction signals, and generate a new WAAS correction signal. This correction signal is then transmitted to INMARSAT geosynchronous communications satellites, which retransmit the correction signal to the entire US. The INMARSAT satellites transmit the correction signal on the GPS L1 frequency, but use a different pseudorandom (PN) code than any of the GPS satellites. The WAAS beacon receiver could be incorporated directly into a GPS receiver. Any user with a receiver designed to read these corrections could remove the error based on information from the most recently uploaded satellite.

The WAAS is based on a network of approximately 25 ground reference stations that covers a very large service area. Installation of the 25th and final WAAS Reference Station (WRS) was completed June 3, 1998 at the FAA Air Route Traffic Control Center in Kansas City, Missouri. WAAS allows a pilot to determine a horizontal and vertical position within 6-7 meters as compared to the 100-meter accuracy available from the basic GPS service.

Satellite-based augmentation systems (SBAS) similar toWAAS, but for other regions of the world are also being developed by Europe (European Geostationary Navigation Overlay System or EGNOS); Japan (Multi-functional Transport Satellite (MTSAT)-based Satellite Augmentation System or MSAS); and India ((GPS Aided Geo Augmented Navigation or GAGAN), and are being considered by additional nations such as Brazil.

Local Area Augmentation System

The second augmentation to the GPS signal is the Local Area Augmentation System (LAAS). The LAAS is intended to complement the WAAS and function together to supply users with seamless satellite based navigation for all phases of flight. In practical terms, this means that at locations where the WAAS is unable to meet existing navigation and landing requirements (such as availability), the LAAS will be used to fulfill those requirements.

Similar to the WAAS concept, which incorporates the use of communication satellites to broadcast a correction message, the LAAS will broadcast its correction message via very high frequency (VHF) radio data link from a ground-based transmitter.

LAAS will yield the extremely high accuracy, availability, and integrity necessary for Category II/III precision approaches. It is fully expected that the end-state configuration will pinpoint the aircraft's position to within one meter or less and at a significant improvement in service flexibility, and user operating costs.

GPS Aiding

GPS Aiding is a form of integrating multiple navigation sources and sensors into the navigation solution. Outside inputs are used to ensure GPS signals are acquired and maintained. In the event GPS signals are lost, these inputs are used to maintain current position until GPS signals can be reacquired. Sources of Aiding include anything that can add to a user's capability to navigate to include Inertial Navigation System (INS).

MILITARY POS/NAV APPLICATIONS

GPS is integrated into DoD combat forces at all levels, from the hand-held receiver carried by the infantryman to the embedded GPS navigation aids on the most modern aircraft to provide precision location determination and navigation support. GPS is a part of the guidance system in most current and all planned precision-guided munitions being acquired by the Services. GPS is also integrated into military forces worldwide, both friend and foe. Forward air controllers, pilots, tank drivers, and ground troops all use GPS to help ensure victory on the battlefield.

GPS receivers can be used for almost any application that requires accurate position location or navigation. It has been said, "If it moves, it can use GPS." There are many different civilian and military GPS receivers that have been developed and new ones are being introduced all the time. In addition to improved stand-alone receivers, GPS is integrated or imbedded into other battlefield systems such as inertial navigation devices, survey equipment, combat net radios, Army Tactical Command and Control System (ATCCS) equipment, and other systems that have a requirement for position determination, velocity or precise time. Even the space shuttle and some satellites use GPS to report their position to ground controllers rather than having to rely on ground-based space surveillance networks.

Space-based Blue Force Tracking

Real time situational awareness of friendly forces is the center piece of the Army's battlefield digitization efforts. The Army has worked hard for years to develop systems capable of tracking forces in the close battle. There had been now no viable joint solution for forward line of troops and deep battle visualization existed.

ARSTRAT has the Space-Based Blue Force Tracking Mission Management Center that provides warfighting combatant commands with near-real time Blue Force Tracking data gathered by space-based systems. The Space based Blue Force Tracking Mission Management Center (SB-BFT MMC) provides Warfighting Combatant Commands with near-real-time (NRT, less than 15-seconds) Blue Force Tracking data gathered by Space-Based systems. This data is pushed as far forward as technically possible and serves as actionable blue force information within the commander and subordinates' Common Operating Picture (COP) and lends to more robust situational awareness and fratricide prevention.

TALON HOOK

After the shoot down of Capt Scott O' Grady in 1995, the TALON HOOK program received special recognition. The USAF created this program to integrate a GPS receiver into the PRC-117 survival radio to take the "search" out of search and rescue. The downed pilot has the HOOK-112 which acts like a regular survival radio, and the pilot can "data burst" transmit his or her position and other information to those who can help. "Those who can help" include:

- National systems that relay information to the processing station and then back to the theater.

- UAVs and U-2s.

- Airborne assets.

- Friendly troops with interrogator units.

Once the location of the downed pilot has been determined and relayed to rescue forces, SAR forces can use the interrogator unit to guide to the pilot's exact location.

Weapon Systems

There are many weapons systems that are or will be using GPS. Aviation systems employ GPS for better position accuracy. The TLAM uses GPS for mid-course guidance, and then switches to terrain navigation for terminal guidance. If terrain navigation fails, GPS guidance is used for the terminal phase. The addition of GPS guidance saves considerable mission planning as well as national systems assets by minimizing the amount of terrain-following navigation required. GPS enables the ATACMS to position itself and calculate the range and firing solution to the target point. The employment of precision munitions like the Joint Direct Attack Munition (JDAM) has produced much less collateral damage. The Joint Precision Airdrop System (JPADS) ensures aircrews get supplies to forces on the ground with less exposure to hostile fire.

Combat Survivor/Evader Locator (CSEL)

Most current survival radios rely primarily on line of sight (LOS) communication. CSEL gives the warfighter the capability to communicate over-the-horizon (OTH) directly with search and rescue forces around the globe via a robust, automated C3 system. The radio incorporates the latest-generation GPS receiver which gives CSEL an unparalleled ability to precisely identify the location of the warfighter.

Artillery Pointing

Artillery batteries can shorten the time needed to survey in guns before they begin operation. This is relatively important in modern warfare because artillery batteries must move often to keep from being hit from counter fire.

CIVILIAN USE OF GPS

The civilian market for GPS receivers is expanding rapidly, in many cases faster than the military market. The scientific community and the aviation community are some of the biggest users of GPS. Receivers can cost $100 for a simple SPS receiver with an accuracy of 100m to $15,000 for a L1/L2 carrier phase survey receiver with accuracy up to 0.01m. Regardless of how many channels and other features to improve performance, all commercially developed sets are C/A-code receivers and are not able to provide PPS because they cannot store the cryptographic codes needed to compensate for SA and cannot decrypt the Y-code.

Civilian and international users have always been concerned with how they are to depend on a system controlled by the US military. GPS is owned and operated by the US Government as a national resource. DoD is the "steward" of GPS, and as such, is responsible to operate the system in accordance with the signal specification. As stated earlier, the March 1996 Presidential Decision Directive, passed into law by Congress in 1998, essentially transferred "ownership" of GPS from DoD to the Interagency GPS Executive Board (IGEB). Now the National Executive Committee for Space-based Positioning, Navigation, and Timing, jointly chaired by the DoD and Transportation, allows for both civil and military interests to be included on all decisions related to the management of GPS. The committee provides management oversight to assure that civil and military needs are properly balanced.

OTHER SPACE BASED NAVIGATION SYSTEMS

GLONASS

The Russian Global Navigation Satellite System (GLONASS) is a counterpart to GPS. Both systems share the same principles in the data transmission and positioning methods. It is based on a constellation of active satellites, which continuously transmit coded signals in two frequency bands, which can be received by users anywhere on the Earth's surface to identify their position and velocity in real time based on ranging measurements. The GLONASS satellites are similar to the GPS satellites except that they do not have on-board atomic clocks. Precise timing is maintained at ground stations and periodically transmitted to the satellites. In spite of being less complex than GPS satellites, GLONASS satellites have demonstrated an operational life of only a few years; therefore the satellites need more frequent replacement than GPS satellites.

In 1982 the first GLONASS satellites were set into orbit, and the experimental work with GLONASS began. Over this time span, the system was tested, and different aspects were improved, including the satellites themselves. Although the initial plans pointed to 1991 for a complete operational system, the deployment of the full constellation of satellites was completed in late 1995, early 1996. The Russians allowed the constellation to degrade; however, since that time, interest in the constellation has been renewed and they have completed launch to attain a complete constellation.

Space Segment

The full space segment of GLONASS would be formed by 24 satellites located on three orbital planes with 8 satellites in each plane instead of the 6 planes of 4 satellites that GPS uses. The three orbital planes are separated by 120 degrees, and the satellites within the same orbit plane by 45 degrees. The GLONASS orbits are roughly circular orbits with an inclination of about 64.8 degrees, a major semi-axis of 25,440 km and a period of 11h 15m 44s. The 64.8 degree inclination gives the Russians better coverage in the northern latitudes than GPS provides.

Control Segment

The ground control segment of GLONASS is entirely located in former Soviet Union territory. The Ground Control Center and Time Standards is in Moscow. The telemetry and tracking stations are in St. Petersburg, Ternopol, Eniseisk, Komsomolsk-na-Amure.

User Segment

The user segment consists of GLONASS receivers that automatically receive navigational signals from at least 4 satellites and measures their pseudoranges and velocities. The receivers simultaneously select and process the navigation message from the satellite signals. The receiver processes all the input data and calculates three coordinates, three components of the velocity vector, and precise time.

A few years ago, the first commercial GLONASS two-channel time receivers became available. More recently, companies have developed new GPS+GLONASS multi-channel and multi-code time receivers. Already a number of major timing centers around the globe observe GPS and GLONASS in multi-channel and multi-code mode.

How GLONASS Works

The coordinate system of the GLONASS satellite orbits is defined according to the FZ-90 system, formerly the Soviet Geodetic System 1985/1990. The time scale is defined as Russian UTC. As a difference from GPS, the GLONASS time system includes also leap seconds.

GLONASS uses Frequency Division Multiple Access to broadcast signals from individual satellites as opposed to the Code Division Multiple Access that GPS uses. This means each GLONASS satellite transmits on a slightly different frequency whereas GPS transmits on the same two frequencies. All satellites transmit simultaneously in two frequency bands to allow the user to correct for ionospheric delays on the transmitted signals. Each GLONASS satellite transmits two types of signals: standard precision (SP) and high precision (HP). SP signal L1 has a frequency division multiple access in the L-band: L1= 1602MHz + n0.5625MHz, where "n" is frequency channel number (n=0,1.2...). This means that each satellite transmits a signal on its own frequency, which is different from the one of the other satellites. These different frequencies allow the user's receivers to identify the satellite. Use of different frequencies for each satellite makes the GLONASS system less vulnerable to interference and jamming than our GPS system.

Superimposed on the carrier frequency, the GLONASS satellites modulate their navigation message. Two modulations can be used for ranging purposes, the Coarse Acquisition code, with a chip length of 586.7 meters and the Precision code, of 58.67 meters. The satellites also transmit information about their ephemeredes, almanac of the entire constellation and correction parameters to the time scale.

Galileo

The European Space Agency and the European Commission have signed contracts to build a European rival to the US GPS. The system, called Galileo, will be compatible with GPS. It is envisioned to consist of at least 21 satellites in medium earth orbit at 15,000 miles and possibly complemented by three geostationary satellites at 22,500 miles, and the associated ground control system. The EU also is looking at charging fees for Galileo's services, or at least any expanded services.

Quasi-Zenith Satellite System (QZSS)

The Japaneses Quasi-Zenith Satellite System [Jun-Ten-Cho in Japanese] is a constellation of at least three satellites, configured such that one of them is always positioned at a high elevation angle over Japan. RF transmission will not be obstructed by tall buildings or mountains, because one of satellites will always remain near high in the sky over Japan at all times. As a result, signal degradation caused by building blockage and multiple signal paths will be less frequent, making the whole system ideal and reliable for mobile data communication and broadcasting. The system is also expected to increase accuracy to GPS users in the eastern Asia area. The service, planned for 2008, can be augmented with the geostationary satellites in Japan's MSAS, currently under development.

Beidou

Although China has not yet established an operational satellite navigation and positioning network, research for such a system has been underway for many years, and a future space-based navigation capability is an acknowledged goal. Beidou ('Big Dipper') is the satellite component for the independent Chinese satellite navigation and positioning system. The Beidou satellite navigation and positioning system are consists of two satellites in geosynchronous orbit. The final Beidou constellation will include four satellites, two operational and two backups. Together with the ground stations, the Beidou system will provide navigation and positioning signals covering the East Asia region. However, to provide global signal coverage, satellites flying in other orbits around the world must complement the system. Three satellites have been launched to date.

POS/NAV CAPABILITIES

GPS provides precise navigation and timing signals for hundreds of operations. It has many capabilities that can be relied on during peacetime and warfighting operations. Capabilities we can expect are:

- GPS is there when need it, in all types of weather, 24 hours a day.

- It is worldwide; the signal is sent directly from the satellite.

- Because the system is completely passive, the unit receiving the global broadcast does not radiate and, therefore, is not susceptible to detection.

- Each receiver set can be set to apply many different datums or grid coordinate systems. Sometimes locations on different maps will have slightly different coordinates. By using GPS to specify the correct datum or grid, those minor "coordinate differences" can be resolved. A GPS receiver can be used to translate coordinates from one datum or grid to another.

- It will coordinate time for all units involved in an operation, allowing coordinated attacks.

- It provides precise time for any equipment needing coordinated timing across the battlefield.

- Because access to the P-code is restricted, only those individuals or units with access can use it.

- The GPS constellation is planned for graceful degradation even if the ground segments are taken out. Graceful degradation means that if for any reason the constellation cannot be commanded, the navigation accuracy will worsen slowly over the course of time - it will not be lost immediately. A Block II vehicle will degrade over 14 days. A Block IIA vehicle will degrade over 67 days. A block IIR vehicle will degrade over 180 days.

- Provides the best position and velocity information available.

- When combined with GLONASS and/or Differential GPS, the timing and location become much more accurate.

GPS accuracy is primarily dependent upon the constellation geometry in relation to the receiver. This geometry can be modeled and used to predict periods of greater and lesser GPS accuracy. Service space professionals use several different programs to produce the GPS accuracy reports for their commands.

POS/NAV LIMITATIONS

Although GPS is a very capable system it also has various limitations. As we become more and more reliant on GPS, it becomes increasingly important to understand its limitations. Knowing the limitations of the system, as well as those of the receiver (which will not be discussed here), the military user can utilize the system to its fullest capabilities without hindering operations. Some factors are limitations of the system, some are limitations of the communications medium and orbit used and some are limitations placed on the system by national policy.

The two most significant problems for military use of space-based radio navigation are the civilian demand for unrestricted, accurate signals and the continued threat of jamming/RFI. Despite the positive force enhancement of space-based radio navigation, there still exists a very grave threat of jamming/RFI to the receivers and associated navigation data links for Precision Guided Munitions

System Limitations

The GPS signal is a low power signal. By the time the signals from the satellites reach the surface of the Earth they are rather weak. The signals cannot penetrate through buildings or metal. Very dense, overhead vegetation can block or weaken the signals. In most forests this has not proven to be a significant problem, however, some problems may exist in dense, triple canopy jungle. An external antenna must be used when using the GPS receiver in a building, vehicle or aircraft. The GPS signal penetrates Kevlar and canvas with very little loss in power. Clouds, rain and snow have little effect; however, extremely heavy rainfall will degrade the signal. Climate, vegetation, and location must all be considered when relying on GPS for missiont planning.

As with all navigation aids, interference, whether intentional or unintentional, is always a concern. GPS signals are communication signals and are subject to the same problems. The signal may be jammed or there may be local interference. The FAA is actively working with the US Department of Defense and other US Government Agencies to detect and mitigate these effects. A number of methods for minimizing interference have been identified and tested and others are being investigated. The FAA is also working to make sure augmentation systems detect and mitigate these effects. The military leader must be aware of the fact that friendly communications systems may interfere with the signal.

Any signal can be jammed. The extremely low power levels of the GPS signals transmitted from space can be overwhelmed with local or mobile jammers. We found this out first-hand during Operation IRAQI FREEDOM in 2003 when Iraqi forces used GPS jammers against our forces in Baghdad, albeit with minimal impact. There are reports of jammers that could jam over a 200-kilometer radius. A company named Aviaconversia announced that it can offer a portable GPS jammer for less than $4000. This jammer is advertised to jam a line-of-sight range of several hundred kilometers. It could realistically be effective at 20 to 40 km.

To maintain its jam resistance, GPS uses two techniques. This combination is difficult, but not impossible to jam. The encrypted GPS frequency on L2 uses a transmission technique called "spread spectrum". The result of using spread spectrum is that one must jam the entire 10 MHz bandwidth to effectively jam the navigation signal. Jamming only part of the bandwidth does not prevent users from receiving and reconstructing the navigation signal. A second method to counter the effects of jamming of the GPS signal is to integrate a GPS receiver into a secondary navigation aid, i.e., combining a GPS receiver and an inertial navigation system (INS).

There is enough concern about GPS jamming that USSTRATCOM has a security program known as Navigation Warfare (NAVWAR). The three principal tenets of NAVWAR are to protect the use of GPS by DoD and allied forces in times of conflict within the theater of operations; prevent the use of GPS by adversary forces; and preserve routine GPS service to all outside the theater of operations. Military users must be aware that the enemy could have the capability to jam the GPS signal at the target location, the friendly locations, or anywhere in the theater AO. Commercial receivers offer no protection in a jamming environment.

Currently the GPS system is dependent on a ground-based control segment to update each satellite in the constellation. The ground-based control system is susceptible to attacks. Once the Block IIR satellites populate the constellation, there will be a cross-link capability so that only one satellite need be updated and in turn, it could update the other satellites in the constellation. This will reduced the reliance on the ground control segment significantly.

The GPS receiver needs to be in the line of sight of at least 4 satellites (hence, it does not work as well in valleys). Most locations will see 6 satellites in line of sight of the receiver but there are still places where the receiver does not have good LOS. Mountainous terrain and urban terrain are among the locations that can cause significant LOS problems. The military user must be aware that line of sight is necessary and consider that fact when using GPS. There are computer programs available that can be used to predict the number of satellites the GPS receiver can see at a given location and at a given time. Using these prediction algorithms can enhance operations that require precise navigation accuracy.

If one is using a differential system, the receiver must also be able to receive the differential signal. The differential signal is different than the satellite signal. The receiver must be designed to receive and utilize both signals.

Communications/Orbit Limitations

The GPS signal is in the UHF spectrum and as such has the same limitations as UHF SATCOM. The UHF frequency range is less jam resistant and is more susceptible to changes in the ionosphere. Interference is also more probable due to the crowded frequency spectrum.

The GPS constellation is located within the Van Allen Radiation Belt. Solar activity (space weather) affects this area more than most layers of space surrounding Earth.

The GPS constellation orbits are in six planes at an approximate inclination of 55 degrees. Although GPS provides worldwide coverage, the coverage in the polar regions is not as good as in more temperate zones. Submarines and aircraft operating at high latitudes may have problems receiving the GPS signals or enough GPS signals to get an accurate position. In addition, the spacing of the satellites in each plane doesn't always give the coverage needed for a specific operation at a specific time. Space tactics have shown what can be done with planned options and creative scheduling techniques to maximize the performance and proper access of the GPS constellation.

Other Limitations

Datum

Although the datum is not a true limitation of the system, it is something that military leaders must concern themselves about when using GPS. A datum is the reference framework for the coordinate system, generally consisting of one or more known positions. A geometric model for the Earth's shape (an ellipsoid or spheroid) is usually associated with the reference framework, to allow the computation of geographic coordinates (latitude & longitude). In the past each country established a datum that best fit its region, based on astronomical calculations. Positions in terms of these local datums may be up to many hundred meters different from positions stated in terms of the global coordinate system used by GPS (the World Geodetic System 1984 - WGS84).

Before starting to collect data with your GPS receiver, one should select the datum that is compatible with the map, chart or digital data are using. By default, GPS uses the WGS84, but many GPS receivers include a menu of local datums that one may select from. The first strategy is preferable if you wish to immediately locate your position on a map or chart, but the second method may be better if you are uncertain of the local datum, or if you wish to later use a better transformation method.

Recognizing Interference

Receivers differ in the way they react to interference. Sometimes, a unit simply ceases to display position information, or maybe the display freezes, or perhaps the device enters a dead-reckoning mode. Whatever the final outcome, and depending on the equipment, users can look for early warning signs of impending failure. A receiver's displayed measure of signal level or signal-to-noise ratio for example, will indicate deterioration in trustworthiness when an interference signal increases in intensity. The number of satellites tracked may also begin to decline. Some receivers will draw the user's attention by whistling or beeping. However, others may give no warning whatsoever. Users should therefore become familiar with the symptoms of possible interference. The most simple way to eliminate direct jamming is t oplace your body between the source of the jamming and your receiver.

PLANNING CONSIDERATIONS

Q. Does the supported command have sufficient Global Positioning Systems (GPS) receivers? In many units there are an inadequate number of GPS receivers to support sustained operations. A secondary consideration is the type of receiver that is distributed within the unit. If you want to use Differential GPS, the receiver must be capable of receiving and using the DGPS signal.

Q. Does the command require a capability to jam or deceive (i.e. spoof) commercial GPS receivers? The ability to deny the use of GPS information to any adversary may be a vital military objective.

Q. Does the threat force have the capability to jam or deceive GPS receivers? The GPS signal is a low-level signal and therefore easily jammed. A broadband jammer can effectively hinder our use of GPS. Multiple low watt jammers could also hinder operations or targeting.

Q. Is the threat using GLONASS? If the threat force is using GLONASS then jamming or denying GPS will not affect their capabilities.

Q. Will the supported command require a GPS differential capability? By using differential techniques, the accuracy of GPS can be significantly improved. This capability can be acquired, but it will require extra equipment, procedures and a fixed location. The GPS receivers must also be capable of receiving the differential signal.

Q. Is there a requirement for knowing "best" GPS coverage times? The satellites in the GPS constellation are constantly moving. Terrain and buildings can mask the signal from the satellites. If specific operations or strike capabilities require accurate GPS, the command must request the capability of identifying those times when the constellation is optimal to support the operation.

Q. What equipment requires the timing signal from GPS? Some communications equipment requires the timing from GPS to maintain frequency hopping and various other operations. GPS jamming or spoofing can affect the timing signal.

Q. Can friendly communications or radar equipment interfere with the GPS signal? Other equipment being used by the friendly forces can jam or interfere with the GPS signal. It is important to know what frequencies all equipment is operating on in order to minimize friendly jamming.

Chapter 8
Environmental Monitoring

WEATHER AND ENVIRONMENTAL MONITORING

Overview

The field of meteorology entered the space age on April 1, 1960 with the launch of TIROS 1 (TIROS stands for Television and infrared Observation Satellite). Some would argue that weather satellites date back to 1959, when NASA's Vanguard 2 photographed the weather on the Earth's surface. Since that time, numerous satellites with ever increasing capabilities and sophistication have been deployed.

Weather satellites provide valuable real-time cloud photographs. Most importantly, coverage includes the 70 percent of the earth's surface covered by water where few surface observations can be made. Before the deployment of weather satellites, many areas had no advance warning of impending severe storms. Today satellites can spot and accurately track hurricanes and typhoons while they are still far out in the ocean. Modern satellites also carry many instruments used to measure various environmental variables, providing vital information to not only meteorologists, but farmers, geologists, fishermen, foresters and others.

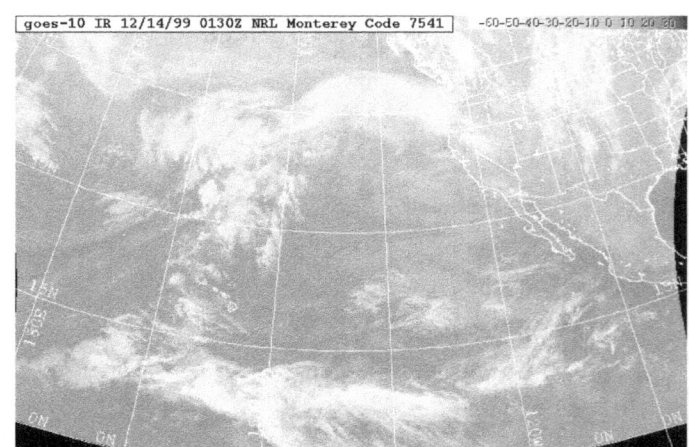

Figure 8-1: GOES-West Image of Pacific Ocean

Satellite based systems do not try to replace surfaced based instruments, but rather to compliment and augment them. Surface based weather reporting instruments are critical to characterizing current weather conditions. Only surface based instruments on land, floating in the oceans, or carried aloft by balloons provide direct precise measurement of environmental conditions. The problem is that there just aren't enough of them. Weather sensors carried on satellites designed to measure current weather conditions and environmental factors that influence the weather are the only alternatives to having many hundreds of thousands of monitoring stations around the world on land and at sea. Obviously weather satellites do not measure environmental conditions directly. Instead, they carry a variety of instruments that either take visible and infrared pictures or take readings, called soundings, of the area within view of the sensors.

Some people refer to weather satellites as environmental satellites because they can tell us something about the air, water and land in which we live. Weather and environmental satellites are similar in that they gather information about the nature and condition of the Earth's land, sea and atmosphere by remote sensing. They accomplish this with sensors that observe the Earth in various discrete bands of the electromagnetic spectrum. They are different in that the systems are designed to observe different phenomena and thus have sensors that gather data in different spectral bands with different resolutions. When the data from space systems is merged with that obtained from other ground and airborne sensors, the resultant products are of significantly better quality than those produced from only one source. For military purposes and for purposes of this text, we will differentiate between weather and environmental satellites. Weather satellites are those designed to assess meteorological phenomena. Environmental satellites are those designed to assess information about the status of the land and changes made over time.

Weather Satellites in General

Most weather satellites are civil systems, even in the case of foreign weather satellites. The United States, Russia, Japan, China, Europe, and India currently have weather satellites in operation. The Defense Meteorological Satellite Program (DMSP) is the only DoD weather satellite system and now managed by the National Oceanographic Atmospheric Administration (NOAA), a civil system. By international agreement the data sent from civil weather satellites is not encrypted and can be received and processed by anyone with the proper type of equipment. Most civil weather satellites transmit data from their sensors in an unencrypted, publicly available format. A number of companies make weather satellite receivers that can receive data directly from the satellites, process it, and display it on a local display. The data rate from some instruments is high so a large antenna may be required to receive all data. To reduce costs, it is common to have central weather satellite receivers with high speed computers to process the data into a variety of weather products. The products are then distributed by radio or over landlines. The organizations operating the civil weather satellites routinely share data. This shared data is available over the Internet to anyone.

The wide-ranging view afforded from satellites makes it possible to observe weather over a large area from an overhead perspective. Weather satellites can be found in two types of orbits: geostationary to give the "big picture" and polar sun-synchronous to collect data closer to the earth. The advantages of using low Earth orbiting satellites and high altitude geostationary orbiting satellites counterbalance the disadvantages of each type of orbit. Using satellites that are in both types of orbits, it is possible to keep 100% of Earth almost constantly under observation and provide the most accurate data available to merge with the ground weather system data.

Various types of sensors are used on board to detect weather phenomena and its changes. Visual sensors on the satellites take pictures of the cloud formations, land and water below. The size and shape of clouds can tell a meteorologist the type of weather in the area of interest. A series of pictures over time can reveal changes in the weather, the speed and direction of movement of storms and other aspects of the weather. Infrared sensors on the satellites can provide digital data on the temperatures of the water, land and clouds in the frame of view. It is even possible to determine the temperature of the atmosphere at various altitudes by a technique called "atmospheric sounding." Sensors on weather satellites can detect the temperature of specific gases that make up the Earth's atmosphere which when merged with other data results in information on wind speed and direction and atmospheric pressure.

Although civil weather satellites have performed extremely well in space for long periods of time, they are not hardened against hostile action or nuclear effects. Only the US military has weather satellites that are hardened to enhance their survivability. They transmit encrypted data which cannot be used by anyone other than approved users who have been provided with the current COMSEC codes. The US military uses weather data provided by the DMSP satellites along with US and foreign civil weather satellites in low Earth polar sun-synchronous and geostationary orbits

MILITARY WEATHER APPLICATIONS

Weather is considered in every facet of military planning, global deployment, and system design and evaluation. In 2003, Peter B. Teets, Undersecretary of the Air Force, testified that "the nation's unparalleled ability to exploit weather and environmental data gathered from space is critical to the success of military operations." With improvements in environmental situational awareness, the US military is rapidly shifting its tactical and strategic focus from "coping with weather" to anticipating and exploiting atmospheric and space environmental conditions for military advantage.

Analysis of weather is a critical step in the Joint Intelligence Preparation of the Operational Environment (JIPOE). Weather and its effects on the environment and terrain have impacted the conduct of military operations throughout history. Knowledge of the current weather and terrain in an area of operations along with accurate predictions of what future conditions is a definite advantage. Timely and accurate knowledge of weather conditions is of extreme importance in the planning and execution of military operations. Real-time night and day observations of current weather conditions provide the field commander with greater flexibility in the use of resources for imminent or ongoing military operations. The military has firmly established the importance of meteorological data from satellites in the effective and efficient conduct of military operations.

Weather data plays a crucial role in planning and conducting all military operations ranging from locating cloud gaps for aerial refueling to determining site visibility prior to reconnaissance flights. We use satellite weather pictures for forecasting when weather conditions will be favorable for friendly forces and the impact of weather on the effectiveness of weapons and the mobility of forces. Cloud cover data are needed to determine weather conditions in data denied and data-sparse regions and to forecast target area weather, theater weather, en route weather (including refueling areas) and recovery weather.

Precipitation information (type, rate) is required to forecast soil moisture, soil trafficability, river stages and flooding conditions that could impact troop and force deployment/employment. Ocean tides information is vital to naval operations for the safe passage in and out of ports, river entrances and for the landing of amphibious craft. Sea ice conditions can have a significant impact on surface/subsurface ship operations. The location of open water areas or areas of thin ice are crucial to submarine surfacing operations, submarine missile launch and penetration by air dropped sonobuoys, which are used for detecting submarines. Knowledge of the location and size of icebergs is also imperative for the safe navigation of surface ships and submarines. This information could provide an important advantage over adversaries in submarine and antisubmarine warfare.

Surface and upper-level wind data are used to support all aspects of military operations, such as assessing radioactive fallout conditions, nuclear, biological and chemical weapon effects, movement of weather systems and predicting winds for weapons delivery. These data are required for all aspects of forecasting support to aircraft and paradrop operations.

Satellite-based remote sensors provide situational awareness of environmental conditions and allow geographical access to areas that otherwise would not be directly available. Weather satellites cover areas that the military does not have direct access to. Weather data from these areas can be critical. For example, weather conditions developing in and around China have a strong influence on the weather that will occur on the Korean peninsula. Although China currently reports some weather conditions from the area through international civil weather reporting channels, the reports may be discontinued during a conflict in Korea. Satellites give us a way to gather weather data quickly over large areas using our own sensors but without violating a country's airspace.

Part of the success of the air campaign in Operation Iraqi Freedom was attributed largely to good weather (for aircraft operations) throughout the period. However, nearly 65% of all air sorties that were cancelled were due to weather during a 3-day period at the end of March 2003.

The ground war commenced on March 20, 2003, and the Third Infantry Division began its furious race through the desert toward Baghdad. As a front swept east across the Mediterranean, forecasters warned to prepare for "the mother of all fronts."

The largest sandstorm to hit southern Iraq in decades engulfed a 300-mile-wide area and blasted tremendous walls of dust into the atmosphere. Meanwhile, the Saddam Fedayeen (Saddam's "Men of Sacrifice") used the cover of the blinding storm to attack the stalled Army convoys.

The same system that blinded troops in southern Iraq created a different set of weather challenges for operations in northern Iraq. Sleet, snow, and heavy cloud cover over Bashur Airfield jeopardized a planned combat jump.

SATELLITE SYSTEMS

Weather is important to all countries and several have satellites in either polar, geostationary or both types of orbits. The United Nations formed the World Meteorological Organization (WMO) in 1951. The purposes of WMO are to facilitate international cooperation in the establishment of networks of stations for making meteorological, hydrological and other observations; and to promote the rapid exchange of meteorological information, the standardization of meteorological observations and the uniform publication of observations and statistics.

The DMSP and NOAA Polar Orbiting Environmental Satellites (POES) have evolved from the first TIROS weather satellites. NOAA also operates the Geostationary Orbiting Environmental Satellites (GOES). The European Meteorological Satellite (EUMETSAT) organization, a consortium of European civil weather reporting agencies, operates the METEOSAT satellites. Japan operates the Multi-Functional Transport Satellite -1R (MTSAT-1R). Russia operates the Geostationary Orbiting Meteorological Satellite (GOMS). India operates the constellation of Indian Satellites (INSAT). INSAT satellites have a multipurpose mission to provide communications and weather data and China operates the Feng Yun low Earth orbiting weather satellites.

Polar Orbiting Satellites

Sun-synchronous, polar orbiting, low Earth orbiting (LEO) weather satellites provide daily, full world coverage and higher resolution imagery than that available from geostationary satellites as polar satellites have the advantage of photographing clouds directly beneath them. Geostationary satellite images of the Polar Regions are distorted because of the low angle the satellite sees the region. Polar satellites also circle at a much lower altitude (about 530 mi., 850 km) providing more detailed information about violent storms and cloud systems. Currently, the United States, China, and Europe operate polar, low Earth orbiting, sun synchronous weather satellites.

The US funds two polar, LEO, sun-synchronous satellite weather systems operated by NOAA:

• Defense Meteorological Satellite Program (DMSP). DMSP provides weather data through all levels of conflict and disseminates global visible and IR cloud data and other specialized meteorological and oceanographic data to support DoD operations.

• NOAA Polar Operational Environmental Satellite (POES) satellites are able to collect global data on a daily basis for a variety of land, ocean, and atmospheric applications. Data from the POES series supports a broad range of environmental monitoring applications including weather analysis and forecasting, climate research and prediction, global sea surface temperature measurements, atmospheric soundings of temperature and humidity, ocean dynamics research, volcanic eruption monitoring, forest fire detection, global vegetation analysis, search and rescue, and many other applications.

Defense Meteorological Satellite Program (DMSP)

Space Segment

The DMSP mission is to generate terrestrial and space weather data for operational forces worldwide. The DMSP satellites design aims at meeting unique military requirements for worldwide weather information. Through these satellites, military weather forecasters can detect developing patterns of weather and track existing weather systems over remote areas. DMSP has been accomplishing this mission for more than 40 years. In that time, the Air Force has successfully orbited more than 35 satellites.

DMSP satellites are in a sun-synchronous orbit at an altitude of approximately 833 km (450 nautical miles or 518 statute miles) with an approximate inclination of 98.7°. At all times at least two operational DMSP satellites are on orbit. Each satellite crosses any point on earth up to two times a day and has an orbital period of about 101 minutes. Their particular orbits allow one satellite to pass overhead in the early morning and the other to pass in the late morning. The satellite encrypts all data transmitted, except when over the north and south poles.

Figure 8-2: DMSP

Each DMSP satellite monitors the atmospheric, oceanographic and solar-geophysical environment of the Earth. The visible and infrared sensors collect images of global cloud distribution across a 3,000 km swath during both daytime and nighttime. The coverage of the microwave imager and sounders are one-half the visible and infrared sensors coverage, thus they cover the Polar Regions above 60° twice daily but the equatorial region once daily. The space environmental sensors record "along track" plasma densities, velocities compositions, and drifts.

The last three Block 5D-2 satellite act as back ups to the two primary satellites. The primary sensor on-board is the Operational Linescan System that observes clouds via visible and infrared imagery for use on worldwide forecasts. A second important sensor is the Special Sensor Microwave Imager (SSM/I), which provides all-weather capability for worldwide tactical operations and is particularly useful in classifying and forecasting sever thunderstorms, hurricanes, and typhoons. Additionally, the SSM/I helps Army operations by getting data on soil moisture, land surface characteristics and vegetation type. The DMSP satellites also measure local charged particles and electromagnetic fields to assess the impact of the ionosphere on surveillance, detection and communications (HF and UHF) systems. Additionally, space weather forecasters use this data to monitor global auroral activity and to predict the effects of the space environment on military satellite operations.

Control Segment

NOAA's Office of Satellite Operations in the National Environmental Satellite, Data and Information Service exercises primary command and control of the DMSP constellation. In May 1994, the President of the United States ordered the Departments of Commerce (DOC) and Defense (DoD) to converge their current polar-orbiting environmental satellite systems into a single National program. Under this directive satellite control authority passed from Air Force Space Command to the tri-service (DoD, DOC, NASA) National Polar-orbiting Operational Environmental Satellite System (NPOESS) Integrated Program Office (IPO). As part of the agreement the DoD transferred day-to-day operations of the DMSP constellation to DOC's NOAA. However, the DoD kept the right to transfer C^2 back to the Air Force in case of extreme national emergencies and exercises oversight through the NPOESS

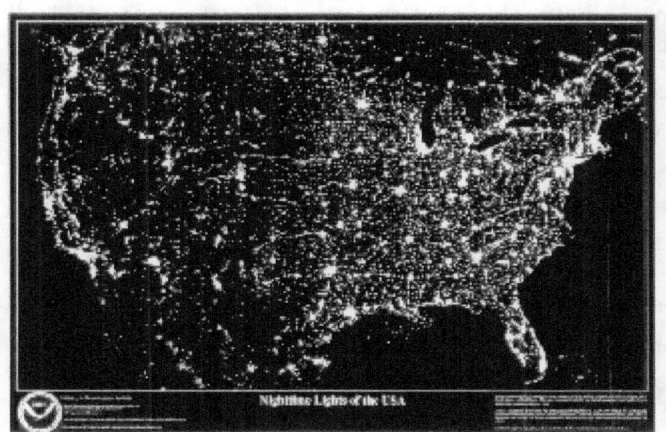

Figure 8-3: DMSP "Night Shot"

IPO's Directorate of Operations. Additionally, in October 1998, the Air Force Reserves activated the 6th Satellite Operations Squadron (6 SOPS) at Schriever Air Force Base, Colorado to act as a hot backup to the DOC Satellite Operational Control Center (SOCC) in Suitland, Maryland. SOCC and 6 SOPS routinely alternate primary C2 responsibilities to maintain proficiency in DMSP operations.

The control segment makes use of ground stations for command and control of the satellites. There are four DMSP-enhanced ground stations in Thule AB, Greenland; New Boston AS, New Hampshire, Kaena Point, HI; and Fairbanks, AK. These sites collect and route environmental data to the DMSP user community. Through specialized communications equipment, the control segment provides all functions necessary to maintain the state of health of the DMSP satellites and to recover the payload data acquired during the satellite orbit. Operators access stored data once every orbit when the satellite is within the Field-of-View (FOV) of a DMSP compatible ground station.

User Segment

A long list of users comprises the user segment of the DMSP constellation. The biggest DMSP data users are the Air Force Weather Agency (AFWA) and the Navy's Fleet Numerical Oceanography Center (FLENUMOCEANCEN). These two agencies serve as the central distribution points for much of the DSMP generated environmental data. AFWA products include: meteorological advice, aviation, terminal and target forecasts prediction of sever weather; automated flight planning; exercise and special mission support; and computations for ballistic missile systems, as well as the collection and dissemination of environmental data. FLENUMOCEANCEN receives and processes DMSP visible, infrared and microwave imagery and distributes products to the Navy's operational forecasting community on-shore and afloat.

Figure 8-5: IR Picture of Hurricane Bonnie

The DMSP User Segment also has a tactical component. This component consists of fixed and mobile land and ship-based tactical terminals operated by Air Force, Navy and Marine Corps personnel. The Air Force provides all meteorological support to the Army. These terminals are capable of recovering direct readouts of real-time visible, microwave and IR cloud cover data from the satellites. The biggest advantage of tactical users having their own receivers and processors is that they can concentrate on their particular areas of interest. Tactical receiver terminals do not, however, receive all of the data received at AFWA or FLENUMOCEANCEN because the satellites transmit it at such a wide bandwidth and at such a high data rate that the size of the terminals would be prohibitively large.

The tactical terminals (TACTERM) have been part of DMSP since the early 1970s. The latest TACTERM is the Mark IV terminal, a transportable satellite terminal designed for worldwide tactical deployment in hostile environments. Mounted on standard shelter, the Mark IV can be towed over virtually any terrain or transported on C-130 or C-17 aircraft and be operational within 8-10 hours. The AN/SMQ-10 and AN/SMQ-11 shipboard receiving terminals are complete satellite meteorological terminals that receive, process, and display real-time DMSP data. The system is designed to be used aboard aircraft carriers and designated capital ships. The SMQ-11, and upgrade to the SMQ-10, is capable of receiving full resolution DMSP OLS and SSM/I data as well as data from other civilian satellites. Additionally, DMSP satellites provide environmental data directly in real time to Air Force, Army, Navy and Marine Corps tactical ground stations and Navy ships worldwide.

IMETS is a mobile, tactical automated weather data receiving, processing and dissemination system designed to provide timely weather and environmental effects forecasts, observations, and decision aid information to multiple command elements at echelons where US Air Force Weather teams provide weather support to the Army. The IMETS is an Army furnished system (standard shelter/vehicle, ACCS common hardware/software and communications) is operated by Air Force Staff Weather Officer Personnel and maintained within planned Army support for system hardware components. The system is capable of receiving weather data from all available sources such as weather satellites, local and remote weather sensors, meteorology; theater forecast units, and Air Force Global Weather Central

The STT (Small Tactical Terminal) is a two person portable satellite data receiver and analysis system. It receives visual and infrared imagery and mission sensor data directly from the satellites. A Basic STT can receive up to 3 streams of satellite data simultaneously. Allows RDS data from DMSP, APT data from NOAA satellites, and WEFAX from geostationary weather satellites to be received and processed at the same time.

NOAA POES

Space Segment

The POES Program is a cooperative effort between NASA and the National Oceanic and Atmospheric Administration (NOAA), the United Kingdom (UK), and France. The Goddard Space Flight Center (GSFC) is responsible for the construction, integration and launch of NOAA satellites. Operational control of the spacecraft is turned over to NOAA after it is checked out on orbit, normally 21 days after launch. The NOAA satellites carry seven scientific instruments and two for Search and Rescue.

The active POES mission is composed of two polar orbiting satellites part of the Advanced Television Infrared Observation Satellites (TIROS) - N (ATN). NOAA-17 (sun-synchronous "morning" orbit) and NOAA-18 (sun-synchronous "afternoon" orbit) operate as a pair and primarily provide data used for long-range weather forecasting ensuring that infrared and non-visible data for any region of the Earth are no more than six hours old. NOAA-15, NOAA-16, and NOAA-18 operate as on orbit back-up.

POES data provides economic, humanistic, and environmental benefits on a continuous, reliable basis. The benefits that directly enhance the quality of human life and protection of Earth's environment include:

• Over 50% of the US public utilizes 3-to-5 day weather forecasts for planning recreational and business activities.

• City, state and federal government agencies utilize TIROS data products to manage resources, plan civic and industrial expansion, schedule services, and monitor population growth.

• Countless lives and properties have been saved by monitoring severe storm movement and forecasting national disasters.

• From monitoring ozone levels and animal migrations patterns to forecasting and detecting forest fires, TIROS is a vital tool of environmental research and protection.

• Global data collected about the earth is used to monitor the environment and trend changes over time.

- Search and Rescue instruments carried on POES satellites contributed to saving over 17,000 lives.

The satellites are a three-axis-stabilized spacecraft that are launched into an 830-870 km, circular, near-polar, sun synchronous orbit. The circular orbit permits uniform data acquisition by the satellite and efficient command and control of the satellite from ground stations located near Fairbanks, AK, and Wallops Island, VA. The constellation consists of two satellites in sun-synchronous orbits. One satellite crosses the equator at 7:30 a.m. local time, the other at 1:40 p.m. local time. Each satellite orbits the Earth 14.1 times per day. Operating as pair, these satellites ensure that data for any region of the Earth are no more than six hours old.

The NOAA satellites are built using the same basic design as DMSP satellites, therefore, they look alike. Sensors carried on the NOAA satellites have different names and different technical characteristics but they perform the same general functions as those on DMSP satellites. The suite of instruments is able to measure many parameters of the Earth's atmosphere, its surface, cloud cover, incoming solar protons, positive ions, electron-flux density, and the energy spectrum at the satellite altitude. As a part of their mission, the satellites can receive, process, and retransmit data from Search and Rescue beacon transmitters, and automatic data collection platforms on land, ocean buoys, or aboard free-floating balloons. The primary instrument aboard the satellite is the Advanced Very High Resolution Radiometer or AVHRR which is similar to DMSP's OLS.

The latest POES satellite, NOAA-19, launched in February 09, is the second in a series of polar-orbiting satellites to be part of a joint cooperation project with the European Organisation for the Exploitation of Meteorological Satellites (EUMESTAT). This satellite works in conjunction with NOAA-17 but also MetOp-A as a primary "morning" pass.

Control Segment

NOAA/NESDIS operates two Command and Data Acquisition (CDA) stations, one in Wallops Island, Virginia and one in Fairbanks, Alaska (formerly Gilmore Creek before 1984), to receive both recorded and direct readout environmental data from the satellite and send these data to Suitland, Maryland, via satellite relay. A receive only CDA station is also set up in Lannion, France at the Centre National d'Etudes Spatiales (CNES), France's national space center. The satellites send more than 16,000 global measurements daily via NOAA's CDA station to NOAA computers, adding valuable information for forecasting models, especially for remote ocean areas, where conventional data are lacking.

User Segment

Some commercial weather receivers can receive POES transmissions. The Army uses POES data for both current weather determination and forecasting to support operations. The POES satellites provide data which enhances data that is provided by geostationary weather satellites. They also provide weather data of the polar areas not covered by geostationary satellites. It is also possible to receive data from civil weather satellites that are operated by other countries.

Figure 8-6: MetOp-A

MetOP (Europe)

MetOp-A is Europe's first operational polar-orbiting weather satellite. Under a new system, MetOp-A shares a common set of core instruments with polar-orbiting meteorological satellites operated by the National Oceanic and Atmospheric Administration (NOAA) in the United States. The MetOp payload includes tried and tested instruments from the US with innovative technology developed in Europe. The satellite is monitored and controlled from a ground station on the island of Spitsbergen in Svalbard, Norway. MetOp-A can measure temperature and humidity, ocean surface wind speed and direction as well as concentrations of ozone and other trace gases.

The satellite includes a data relay system, linking up to buoys and other data collection devices.

Feng Yun (China)

The Feng Yun 1 is a Chinese weather satellite in a sun-synchronous polar orbit with an altitude of 849 km and an inclination of 98.7 degrees. The satellite has four sensors in the visible light band and one in the infrared band. The highest resolution is about 1 km. Each image covers an area about 1,600 km wide (east-west) and 3,200 km long (north-south). It continuously transmits data in analog format for Automatic Picture Transmission (APT) and in digital format for High Resolution Picture Transmission (HRPT). These signal formats are compatible with US NOAA satellites; therefore, receivers capable of receiving NOAA data can also receive Feng Yun data.

A Feng Yun 1C was launched in May, 1999 at an altitude of 870 Km with an inclination of 98.85 degrees. It carried the Shi Jian 5 research satellite as a secondary payload to study the radiation belts. The Chinese destroyed this satellite by a direct ascent anti-satellite missile in January of 2007.

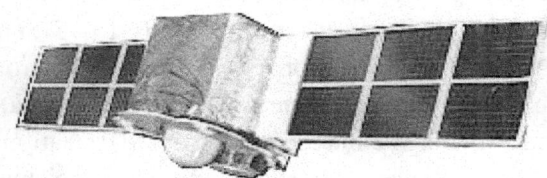

Figure 8-7: Feng Yun 1

The FY-1D is the latest weather satellite in operation having been launched in May 2002. On board the satellite is an instrument called a Multichannel Visible and Infrared Scan Radiometer (MVISR) that has 10 channels. Four of these channels are in the visible region of the electromagnetic spectrum, three in the near infrared, one in the short infrared and 2 in the long infrared. The spatial resolution of the instrument is 1.2 km.

The meteorological satellite data receiving and processing system consists of three ground stations located in Beijing, Guangzhou, Urumqi respectively and a data processing center in NSMC. The system has successfully accomplished the tasks of data receiving, transmitting, processing and product distributing for the FY series of satellites. The system also holds responsibility for receiving and processing the data from NOAA satellites.

China has since updated their polar operating systems with the Feng Yun 3A launched in May 2008 with much better resolution of 250 meter versus the 1.08 Km of the Feng Yun 1-D.

Geostationary Weather Satellites

Another group of weather satellites is maintained in geostationary orbit to provide a continuous watch of the weather on the Earth below them. Each geostationary weather satellite is able to continuously scan approximately one third of the Earth, collecting visual and IR data. This positioning allows continuous monitoring of a specific region. Geostationary satellites measure in "real time", meaning they transmit photographs to the receiving system on the ground as soon as the camera takes the picture. A succession of photographs from these satellites can be displayed in sequence to produce a movie showing cloud movement. This allows forecasters to monitor the progress of large weather systems such as fronts, storms and hurricanes. Wind direction and speed can also be determined by monitoring cloud movement.

The United States, Europe, Japan, China and India operate geosynchronous weather satellites. The system operated by the United States is the Geostationary Orbiting Environmental Satellites (GOES). There are no military geostationary weather satellites.

Geostationary Operational Environmental Satellite (GOES) (US)

Space Segment

Over the past 30 years, environmental service agencies have stated a need for continuous, dependable, timely, and high-quality observations of the Earth and its environment. The average time it takes to get a DMSP/TIROS product to its user is 15-45 minutes depending on satellite overpass and priority of the tasking. The Geostationary Operational Environmental Satellites provide half-hourly observations to fill the need. The instruments on board the satellites measure Earth-emitted and reflected radiation from which atmospheric temperature, winds, moisture, and cloud cover data can be derived.

Figure 8-8: GOES

The GOES constellation is also operated by NOAA. GOES is a series of meteorological geostationary orbiting satellites that provide weather prediction data for the Western Hemisphere and particularly for the US GOES imagery is accessible to over 10,000 ground stations in 120 nations. Because they are in a geostationary orbit, they provide a constant vigil for the atmospheric "triggers" of severe weather conditions such as tornadoes, flash floods, hail storms, and hurricanes. When these conditions develop the GOES satellites are able to monitor storm development and track their movements. GOES satellite imagery is also used to estimate rainfall during the thunderstorms and hurricanes for flash flood warnings, as well as estimates of snowfall accumulations and overall extent of snow cover. Such data helps meteorologists issue winter storm warnings and spring snow melt advisories. Satellite sensors also detect ice fields and map the movements of sea and lake ice. Lastly, they can monitor solar flare activity.

The NASA manages the design, development, and launch of the spacecraft. Once the satellite is launched and checked out, NOAA assumes responsibility for the command and control, data receipt, and product generation and distribution. During NASA's construction and launch phases, the satellites have alphabetical designations: GOES-I, GOES-J, etc. Once the satellites are deployed, they get a serial number in orbit, i.e. GOES-I became GOES-8, GOES-J became GOES-9, GOES-K became GOES-10, etc.

Each satellite in the series carries two major instruments: an Imager and a Sounder. These instruments acquire high resolution visible and infrared data, as well as temperature and moisture profiles of the atmosphere. The three-axis, body stabilized spacecraft design enables the sensors to image clouds, monitor earth's surface temperature and water vapor fields, and sound the atmosphere for its vertical thermal and vapor structures. GOES-8 and GOES-10 also introduce two new features: flexible scanning that allows small-area imaging *plus simultaneous and independent imaging and sounding*, allowing continuous gathering of data from both instruments. They continuously transmit these data to ground terminals where the data are processed for rebroadcast to primary weather services both in the United States and around the world, including the global research community. The processed data are received at the control center and disseminated to the National Weather Service's (NWS) National Meteorological Center, Camp Springs, Maryland, and NWS forecast offices, including the National Hurricane Center, Miami, Florida, and the National Severe Storms Forecast Center, Kansas City, Missouri. Department of Defense installations, universities, and numerous private commercial users also receive processed data. Color is added to indicate areas of severe weather, along with latitude and longitude lines and country outlines so users can relate the weather to the ground below.

Imager

The imager detects different wavelengths of energy through different channels. This allows the imager to capture visible light, emitted long wave radiation and other radiation wavelengths. The imager has five "channels" which monitor radiation at a specific wavelength per given channel. Channel and product descriptions are given below:

0.52 - 0.72 micrometers (visible) - at 1 km, useful for cloud, pollution, and haze detection and severe storm identification.

3.78 - 4.03 micrometers (short wave infrared window) - at 4 km, useful for identifying fog at night, discriminating between water clouds and snow or ice crystal clouds, detecting fires and volcanoes, and determining sea surface temperatures.

6.47 - 7.02 micrometers (upper level water vapor) - at 4 km, useful for estimating regions of mid-level moisture content and advection plus tracking mid-level atmospheric motions.

10.2 - 11.2 micrometers (long wave infrared window) - at 4 km, familiar to most users for cloud-drift winds, severe storm identification, and location of heavy rainfall.

11.5 - 12.5 micrometers (infrared window more sensitive to water vapor) - at 4 km, useful for identification of low-level moisture, determination of sea surface temperature, and detection of airborne dust and volcanic ash.

Sounder

The GOES Sounder is a 19-channel discrete-filter radiometer covering the spectral range from the visible channel wavelengths to 15 microns. It is designed to provide data from which atmospheric temperature and moisture profiles, surface and cloud-top temperatures and pressures, and ozone distribution can be deduced by mathematical analysis. It operates independently of and simultaneously with the imager, using a similarly flexible scanning system. The sounder's multi-element detector array assemblies simultaneously sample four separate fields or atmospheric columns. A rotating filter wheel, which brings spectral filters into the optical path of the detector array, provides the infrared channel definition.

The GOES system performs the following basic functions:

- Acquisition, processing, and dissemination of imaging and sounding data.

- Acquisition and dissemination of Space Environment Monitor (SEM) data.

- Reception and relay of data from ground-based Data Collection Platforms (DCPs) that are situated in carefully selected urban and remote areas to the NOAA Command and Data Acquisition (CDA) station.

- Continuous relay of Weather Facsimile (WEFAX) and other data to users, in dependent of all other functions.

- Relay of distress signals from people, aircraft, or marine vessels to the search and rescue ground stations of the Search and Rescue Satellite Aided Tracking (SARSAT) system. A dedicated search and rescue transponder on board GOES is designed to detect emergency distress signals originating from Earth-based sources. These unique identification signals are normally combined with signals received by a low-Earth orbiting satellite system and relayed to a search and rescue ground terminal.

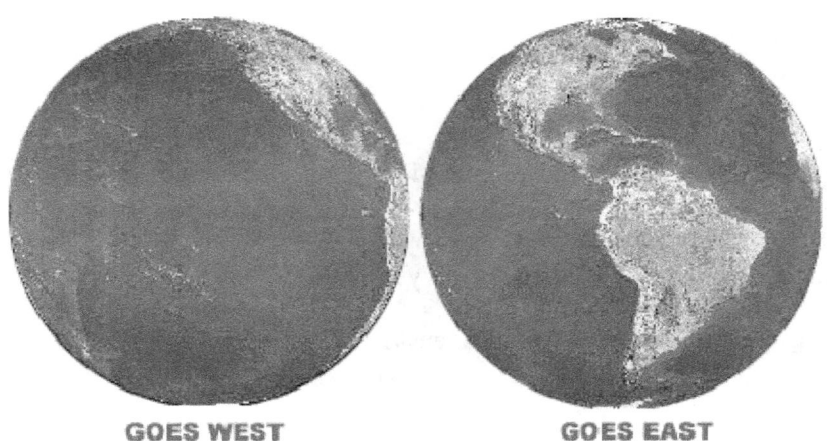

GOES WEST   GOES EAST

Figure 8-9: GOES 10 and 12

The GOES I-M system serves a region covering the central and eastern Pacific Ocean; North, Central, and South America; and the central and western Atlantic Ocean. Pacific coverage includes Hawaii and the Gulf of Alaska. This is accomplished by keeping two satellites in orbit at all times. The current operational constellation consists of GOES-15 (or GOES West) located at 135°W longitude and GOES-13 (or GOES East) at 75°W longitude.. Coverage extends approximately from 20°W longitude to 165°E longitude.

A common ground station, the CDA station located at Wallops, Virginia, supports the interface to both satellites. The NOAA Satellite Operations Control Center (SOCC), in Suitland, Maryland, provides spacecraft scheduling, health and safety monitoring, and engineering analyses.

COSPAS-SARSAT

Although not a weather system, several of the weather satellites carry a package on board to aid Search and Rescues efforts. The COSPAS (Cosmicheskaya Sistyema Poiska Avariynich Sudov) - SARSAT (Search And Rescue Satellite) system is an International, humanitarian satellite-based search and rescue system which can detect and locate transmissions from emergency beacons carried by ships, aircraft, or individuals. It has helped save over 9025 lives worldwide since its inception in 1982. There are 25 other participating nations in the COSPAS-SARSAT program.

Figure 8-10: COSPAS-SARSAT Emblem

The GOES and India Satellite (INSAT-3A) satellites carry a Search and Rescue (SAR) module that listens for emergency distress signals from special transmitters on the ground. There are a few basic models of these emergency transmitters or beacons but the concept for all of them is to transmit a signal that identifies the transmitter.

The GOES geosynchronous altitude vantage point gives the GOES SAR the capability to detect distress signals from one entire hemisphere of Earth 24 hours a day but the GOES satellites are too far away to pinpoint the location. Upon detection of a distress signal, other satellites in LEO with SAR equipment are used to pinpoint the location of the distress signal with an accuracy of one to two km. LEO satellites with SAR systems include US NOAA-15 to 19 and the Russian Nahezda civil navigation satellite constellation.

The COSPAS-SARSAT ground network includes the US, Canada, France, Russia, Norway, and Brazil. The US portion of the COSPAS-SARSAT System is operated by the NOAA SARSAT Office in Suitland, Maryland. Rescue services around the world have responded to save the lives of hundreds of civilians injured in aircraft accidents and distress at sea. Some military aircraft carry COSPAS-SARSAT emergency beacon transmitters for use during peacetime.

METEOSAT (EUMETSAT) (Europe)

METEOSAT weather satellites are owned and operated by EUMETSAT, a consortium of 21 European States (Austria, Belgium, Croatia, Denmark, Finland, France, Germany, Greece, Ireland, Italy, Luxembourg, the Netherlands, Norway, Portugal, Slovak Republic, Slovenia, Spain, Sweden, Switzerland, Turkey and the United Kingdom). EUMETSAT also has nine Cooperating States (Bulgaria, Czech Republic, Estonia, Hungary, Iceland, Latvia, Lithuania, Poland, and Romania). These States fund the EUMETSAT programs and are the principal users of the systems. EUMETSAT took over formal responsibility for the Meteosat system in January 1987 and 1991 had initiated a new program to ensure the continuation of Meteosat operations.

EUMETSAT's Meteosat system is intended primarily to support the National Meteorological Services (NMS) of Member States. The NMS in turn distribute the image data to other end users, notably through the provision of forecasts on television several times a day. Through this particular distribution system it could be said that most of the population of Europe makes direct use of Eumetsat's imagery. Second priority is given to the NMS of non-Member States. These are given privileged access to Meteosat data in the continuing tradition of data exchange between meteorological services. They too use the data for the preparation of forecasts and for distribution to television audiences.

Meteosat-9 is the primary operational spacecraft and positioned in geostationary orbit at about 0° east longitude. Meteosat-8 is at 10° as a back-up. Meteosat-7 has been in operation since June 1998. It is in geostationary orbit above the equator at 57° longitude off the east coast of central Africa for prime coverage of the Indian Ocean.

Meteosat sensors are similar to GOES. The Meteosat system provides continuous and reliable meteorological observations from space to a large user community. In addition to the provision of images of the earth and its atmosphere every half an hour in three spectral channels (Visible, Infrared and Water Vapor) a range of processed meteorological parameters is produced. Meteosat also supports the retransmission of data from data collection platforms in remote locations, at sea and on board aircraft, as well as the dissemination of meteorological information in graphical and text formats.

INSAT (India)

India's INSAT series of geostationary spacecraft perform the dual missions of communications and meteorology. These unique satellites carry telephone, and television transponders along with weather sensors.

INSAT satellites carry a Very High Resolution Radiometer (VHRR) with 2-km resolution in the visible band and 8-km resolution in the IR band. The sensors are similar to those on GOES. Like many GEO meteorological satellites, INSAT spacecraft require 30 minutes to complete a full Earth scan. In addition to full Earth images, the VHRR can be commanded to scan very limited regions for more rapid return of time-critical data, e.g., during the approach of cyclones to the sub-continent. Each vehicle is also capable of receiving (on 402.75 MHz) meteorological, hydrological, and oceanographic data from remote data collection platforms for relay to central Indian processing centers.

The INSAT 2 program was inaugurated in 1992 with the launch of INSAT 2A, followed by INSAT 2B in 1993. INSAT 2 satellites also carry the Data Relay Transponder system for collection and retransmission of data. The current constellation includes INSAT 3A at 93.5E, INSAT 3C with limited capability at 74° E, and Kalpana-1 also at 74°E.

The satellites are controlled from the INSAT Master Control Facility at Hassan, Karnataka, India. In compliance with the international agreement on weather satellites, the data are not encrypted, however, the signals from the satellite are transmitted on a narrow spot beam to the Delhi Earth Station. The spot beam limits reception of the data to a small area. India did not share data from this satellite until 1991 and then the data was 3 years old. The probable reason for the narrow transmission beam and the lack of sharing of the data is to keep certain neighboring countries from getting the information. The data is relayed to the Meteorological Data Utilization Center in New Delhi where the data is processed. The meteorological products are then relayed through an INSAT satellite to 22 Secondary Data Utilization Centers throughout India.

Meteorological Satellite (MTSAT) (Japan)

The MTSAT Program consists of a series of satellites operated by the Japan Meteorological Agency. The satellite is also known as Himawari. MSAT-1R, launched in 2006, is in geostationary orbit at 140°E which allows the satellite to image the Pacific Basin as the standby system and MTSAT-2, launched in 2007, is the operational system at 145°E longitude. The first satellite in the series was launched in 2006 and the last in 2007. The satellites have been used for the World Weather Watch Program.

The MTSAT has an onboard sensor, which is called the Japanese Advanced Meteorological Imager (JAMI). JAMI obtains round earth imagery, called "full-disk image", and observes earth surface conditions and cloud distributions as well as meteorological phenomena such as typhoons, depressions, front and so on. In addition, the various meteorological parameters, such as sea surface temperature and cloud motion winds are extracted from image data.

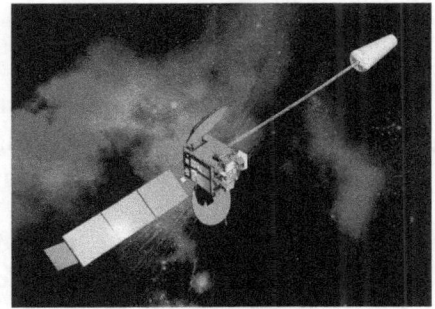

Figure 8-10: MTSAT-1R

MTSAT has communication functions to disseminate observed image data and processed pictures. The original raw data of all channels are processed to normalized geostationary projected images every time, and are disseminated to the users of Medium-scale Data Utilization Station (MDUS) with High Rate Information Transmission (HRIT). After the observation of the earth image and its data processing, the weather facsimile (WEFAX) pictures and The Low Rate Information Transmission (LRIT) image data are disseminated to the users of Small-scale Data Utilization Station (SDUS) via MTSAT-1R, according to the daily operation schedule. The information on the seismic information and Tsunami Warnings are disseminated to the domestic disaster protection authorities using the interrogation function of DCS of MTSAT-1R in order to prevent the natural disaster caused by earthquakes and Tsunamis.

GOMS (Elektro) (Russia)

The year 1994 witnessed the long-awaited debut of the Geostationary Operational Meteorological Satellite (GOMS) system of Elektro spacecraft. GOMS was launched 31 October 1994 and placed into a geostationary orbit at 76.61 degrees E. The GOMS network will eventually consist of three spacecraft spaced 90 degrees apart in the geostationary ring: at 14 degrees W, 76 degrees E, and 166 degrees E. Each 2.6-metric-ton spacecraft will have an estimated operational lifetime of at least three years. The satellites are a 3-axis-stabilized platform.

Onboard instruments package allows:

- obtaining in real time visible and infrared images of the Earth surface and cloud cover within a radius of 60° 50' centered at sub-satellite point;

- providing continuous observation of the dynamics of varying atmospheric processes;

- detecting, on an operational basis, hazardous natural phenomena;

- determining wind velocity and directions at several levels, sea surface temperature;

- obtaining information on fluxes of solar and galactic particles, electromagnetic ultraviolet and X-ray radiation, variations in the vector of magnetic field.

Twelve communications channels link the spacecraft to the receiving and processing centers, the independent data receiving center, and the data collection platforms. The main data receiving and processing center is in the Moscow region while two regional centers are located at Tashkent and Khabarovsk.

FengYun 2 (China)

The Chinese Meteorological Administration (CMA) launched FengYun 2C (FY-2B) to 105 degrees East longitude on 10 Oct 2004. Feng Yun means Wind and Cloud in Chinese. The satellite was located above the equator of 105°E and acquired the first visible image on 1 January 2005.

FY-2D was launched by a CZ-3A booster from Xichang on 8 December 2006. The ground station received the first image from the satellite at 14:00 on 12 January 2007. The satellite was positioned at the 'back-up' position 86.5 degrees E. in geosynchronous orbit, covering most areas of Asia, the Indian Ocean, and the West Pacific. The satellite formed a 'twin-star' weather forecast network with the FY-2C to provide comprehensive weather information.. Feng yun 3A is now at 105 degrees E and completes coverage of the area.

The primary payload on FY-2 is the Visible Infrared Spin-Scan Radiometer (VISSR); an imaging instrument consists of a scanning system, a telescope, and infrared and visible sensors. The spectral coverage includes the three conventional channels in the wavelength bands of visible (0.55-1.05 micrometer), infrared (10.5-12.5 micrometer) and water vapor (6.2-7.6 micrometer). Using sensors to sense radiant and reflected solar energy from sampled areas of the Earth, VISSR can make daytime and nighttime observations of cloud, and determine cloud heights, temperatures and wind fields. FY-2 produces a cloud image once every half an hour, with a nadir resolution of 1.25 km in the visible channel. Nominally FY-2 transmits 28 cloud images daily. During the flood season, FY-2 can increase the transmission to 48 or more images per day. The satellite also carries instruments to monitor solar activities such as emission of x-rays, and measure particle radiation in the orbital environment.

NSMC, a scientific research and operational facility affiliated to the China Meteorological Administration (CMA), receives, processes, and distributes metsat data to users. FY-2 climatic information of western Asia and the Indian Ocean will be distributed to the international community.

FY-2 will continue the surveillance of changing weather conditions that spans the Indian Ocean in the west and the western Pacific Ocean in the east. The metsat monitor development and movement of typhoons in the Pacific Ocean and cyclones in the Indian Ocean, watch weather changes at the Tibetan Plateau, survey the land and seas, detect grassland and forest fires, and observe sandstorms and fog formation.

WEATHER SUPPORT

Air Force Weather Agency was formed on Oct. 15, 1997, as part of a reengineering effort to streamline and improve the structure of the former Air Weather Service. This was a result of the realignment of Air Weather Service headquarters staff from Scott AFB, Ill. and the former Air Force Global Weather Center, DOD's primary centralized weather production facility at Offutt.

The Air Force Weather Agency mission is to enhance our nation's combat capability by arming our forces with quality weather and space products, training, equipment and communications -- anytime, anywhere. AFWA's production operation involves gathering over 140,000 weather reports per day from conventional meteorological sources throughout the world and relaying them to AFWA by the Automated Weather Network (AWN). By combining these data with information available from military and civilian meteorological satellites, AFWA constructs a real-time, integrated environmental database. A series of scientific computer programs model the existing atmosphere and project changes. AFWA is responsible for providing technical advice and meteorological assistance to Air Force weather units supporting Active or RC Army units.

AFWA exchanges data and meteorological products with the National Weather Service and the Naval Oceanography Command. AFWA is the backup agency for two National Weather Service centers. Support to the National Meteorological Center includes products transmitted on the Digital Facsimile (DIFAX) circuit and aviation winds for civilian users. Support to the National Severe Storms Forecast Center includes severe weather forecasts to the civilian community. Products and services provided by AFWA include meteorological advice; aviation, terminal and target forecasts; prediction of severe weather; automated flight planning; exercise and special mission support; and computations for ballistic missile systems, as well as the collection and dissemination of environmental data.

There are a number of ways to get weather satellite support. The DMSP terminals discussed can receive a direct downlink although they will not receive the same amounts of information that AFWA receives. Users can dial-into AFWA and FNMOC. It is possible to set up an account and dial in, using a modem over a telephone line. A SATCOM link can be established to AFWA to receive The Automated Weather Dissemination System (TAWDS) broadcasts. There are many sites on the World Wide Web that provide weather information from around the world. Almost all of it is processed data converted into an image or low resolution data. Some sites give weather information for specific cities or regions. Most homepages are operated by civilian agencies or even civilian individual citizens. Lastly, special support can be coordinated by the theater staff meteorologist.

WEATHER SATELLITE CAPABILITIES

A prime advantage of environmental satellites is their ability to gather data regarding remote or hostile areas, where little or no data can be obtained via surface reporting stations. Weather satellites rely on gathering data in the visual, infrared, and microwave spectral bands. Infrared sensors provide images which are based on thermal characteristics of atmospheric features, such as clouds, and earth features, such as land masses and water bodies. This data can be used to calculate the altitude of cloud tops and ground or water surface temperatures. This data can be gathered during light or dark hours.

Thermal and visible images together provide the coverage and extent of clouds at various levels, as well as other physical phenomena such as ice fields and snow. Microwave sensors are used to measure or infer sea surface winds, ground moisture, rainfall rates, ice characteristics, atmospheric temperatures, and water vapor profiles.

Weather satellites can gather information about the magnetosphere and ionization of the atmosphere. This data can be used by the space weather forecasters to predict communication outages and to alert operators to look for possible satellite problems.

Current polar-orbiting satellites provide high resolution imagery and can image weather over the poles. Geosynchronous satellites provide a constant look at the same area with an updated weather picture every 30 minutes. Geosynchronous satellites have a large field of view and are utilized to track large weather systems and provide environmental warning.

WEATHER SATELLITE LIMITATIONS

Weather satellites have a limited multi-spectral capability. Some meteorological parameters needed by forecasters for operational support cannot currently be accurately determined from satellites, including, heights of cloud bases, visibility restrictions, and lower level winds.

Current polar-orbiting satellites have limited data-refresh rates. A constellation with two satellites will give a picture of the same area at least every 6 hours (except for the poles which get covered more often). Geostationary satellites provide lower resolution images. The image quality degrades as distance and angle from the point directly under the satellite increases. The resolve of an increased angle is poor coverage as polar latitudes are approached and no coverage at the poles.

PLANNING CONSIDERATIONS

Understanding the capabilities and limitations of weather satellites and the orbits they are in is crucial to effective operational planning. Weather affects all operations in one way or another. The weather data obtained from ground systems is effective in determining the current situation. Satellite data is necessary for accurate forecasting. Satellite data from polar, sun-synchronous orbiters is different than data from geostationary orbiters.

Typically, weather satellite information is broadcast worldwide. Anyone with a PC and Internet connection can obtain a current weather picture. What we see is what the enemy also can see. DMSP is the only encrypted weather signal and therefore its information can be denied to the enemy. However, there are instances now and there could be instances in the future when weather data is denied. India has denied its real-time weather information in the past. Meteosat has encrypted its data and then sold access to the data to secondary and other partners. An example of the effects of denying weather data would be in a hostile engagement in Korea. The weather patterns over China greatly affect the future weather over Korea. China could deny its weather data. If India and Russia do the same, we would have a gap in coverage that could lead to inaccurate forecasting.

Q. What quality and frequency of satellite imagery, forecast products and communications equipment are required? Are civilian weather satellites adequate or is there a need for encrypted data from military satellites or a mix of both? Is "CNN" type weather forecasting adequate? The frequency and quality of required weather forecasts will determine what kind of receive equipment need.

Q. Does the command require the capability to prevent an adversary from receiving meteorological transmissions? Due to treaty limitations, political considerations, and the impact on US and friendly forces, denial of transmissions may not be a desirable option. Targeting enemy earth terminals remains an option.

Q. Are the current available transmission times for DMSP and civil meteorological satellites adequate? In a highly maneuverable, fast-paced conflict, timely weather data is important. Current military weather systems do not provide full-time coverage. However, a mix of military and civil assets can provide increased coverage.

Q. Does the supported command require rapid or near-real-time access to recorded DMSP data? Microwave imagery data is available in a recorded format, and through a real-time direct downlink. ARSTRAT can provide the necessary equipment for direct downlink. The Air Force can provide the weather analyst. Prior coordination for both is necessary.

Chapter 9
Early Warning

SPACE-BASED MISSILE WARNING

Theater Ballistic Missile (TBM) Detection and Warning:

Theater Ballistic Missile proliferation is becoming an ever-increasing problem in the world in the 21st Century. Many nations possess theater ballistic missiles and some have made this technology available for purchase. Today, proliferation poses a significant threat to US field commanders in overseas locations and this threat will continue to grow in the future. Detection and warning of enemy ballistic missile launches allow commanders to take appropriate passive missile defense, attack operations, and active missile defense actions. TBM warning is a subset of Surveillance and involves the sensors (Defense Satellite Program (DSP) and Space Based Infrared System (SBIRS)), communication links and verbal command directions. These satellite sensors also accomplish nuclear detonation (NUDET) detection and can be used to detect other IR events. Ground stations such as Joint Tactical Ground Stations (JTAGS) and the Space Based Infrared System (SBIRS) Mission Control Center gather, process and disseminate early warning information.

Theater Ballistic Missile Warning

One of the primary missions of the United State Strategic Command is to provide space-based theater ballistic missile warning to US forces worldwide. This warning provides the troops in the field the opportunity to defend themselves or take the necessary precautions in the event of a missile threat. Those precautions could include intercepting the missile when combined with the current and future theater missile defense systems and the evacuation of buildings in the threatened area. The command performs this mission with a variety of ground-based and space-based systems as part of the Theater Event System (TES).

The Theater Event System consists of the DSP satellite constellation, SBIRS, JTAGS, and Tactical Detection and Reporting System (TACDAR). The data from these sources is disseminated world wide via Integrated Broadcast Service (IBS).

A brief description of key Theater Event Systems and sensors follows.

Defense Support Program (DSP)

DSP satellites are a key part of North America's early warning system. In 22,000 mile geosynchronous orbits, DSP satellites serve as the continent's first line of defense against ballistic missile attack and are often the first system to detect world-wide missile launches. DSP satellites use an infrared sensor to detect heat from missile or booster plumes against the relatively cool background of the Earth's surface. These satellites have provided uninterrupted warning since the early 1970s when they were first launched. These satellites were designed to detect strategic ballistic missiles in the early stage of launch of their flights. However,

Figure 9-1: DSP Satellite

during Desert Storm, prior to the start of the air campaign, the detection software was upgraded and refined to detect short-range theater ballistic missiles such as the SCUD missile. In addition to missile launches, the DSP system also has numerous sensors on board to detect nuclear detonations.

During Operation DESERT STORM, a DSP satellite was able to detect the launch of Iraqi SCUD missiles toward Saudi Arabia and Israel. The satellite sensor data was transmitted to a CONUS processing station. Computers analyzed the data to determine when a launch occurred. If the operator confirmed the computer analysis, a launch detection alert was issued. This alert message was relayed over satellite communications to the headquarters in Saudi Arabia. The alert message provided early warning to military and civilian personnel in the target area and provided cueing information to the Patriot missile batteries providing point defense. This procedure was relatively expedient for its time. It was in fact using what had been designed as a strategic missile warning system via Cheyenne Mountain and NORAD for tactical purposes. It was used operationally for the first time during Operation DESERT SHIELD/STORM. The TES was created post DESERT STORM to allow for better, timelier missile warning data to flow to theater. TES now allows for theater commanders to take action without having to wait for NORAD's North American threat assessment.

While DSP proved very effective, the fact remained that it had been designed for strategic missile warning. In response, the program office launched Talon Shield Phase 1, which quickly fielded an operational system known as Attack and Launch Early Reporting to Theater (ALERT) in March 1995. ALERT was a high-confidence operational system that provided assured theater missile warning to warfighters worldwide. ALERT was deactivated in September of 2002 and the MCS team now performs the ALERT mission.

Space Segment

The DSP program came to life with the first launch of a DSP satellite in the early 1970s. Since that time, DSP satellites have provided an uninterrupted early warning capability that has helped deter superpower conflict.

The DSP satellites are launched into a geostationary orbit from which the sensors on the satellite monitor the Earth below for the launch of ballistic missiles.

The primary mission of DSP is to detect, characterize and report in real time, missile and space launches occurring in the satellite Field Of View. DSP satellites track missiles by observing infrared (IR) radiation emitted by the rocket's exhaust plume. The principal sensor subsystem is the Infrared (IR) Telescope. Infrared energy given off by hot sources on the Earth is detected by an array of photoelectric cells located in the IR Telescope. Sensor data is transmitted to control segment ground stations for processing. The DSP satellites also carry RADEC sensors capable of detecting and quantifying nuclear explosions on the Earth's surface, in the atmosphere and in near Earth space.

The sensor and the spacecraft, which together comprise the satellite, are placed in geosynchronous-equatorial orbit so that the telescope is pointed toward the earth and rotated at six revolutions per minute. To provide a scanning motion for the infrared (IR) sensor, the satellite is spun about its Earth-pointing axis.

Over the last 30 years, there have been 23 satellite launches with five major design changes. The last DSP satellite, DSP Flight 23, was launched on an EELV during the second half of 2006.

The follow-on to the DSP is SBIRS.

Control Segment

The DSP satellites are operated and controlled by the Air Force Space Command's 460th Space Wing's 2nd Space Warning Squadron located at Buckley AFB, Colorado. DSP ground support consists of a Mobile Ground System (MGS) consisting of six fully deployable units (tractor-trailer rigs) and the fixed-site SBIRS Mission Control Station (MCS) that receives, processes, and reports mission data to the users.

The SBIRS MCS represents a transformational step in the evolution of the nation's space-based infrared systems. The MCS centralizes global command, control, and communications for strategic and tactical warning into a single modern peacetime facility. Emerging from a heritage of over 30 years of early warning and the lessons of the 1991 war with Iraq, the consolidated facility provides warfighters with timely, unambiguous missile warning reports.

The MCS operates the DSP satellites today and will have the capability to operate the SBIRS constellation from the consolidated location in the future.

The first step toward a more robust infrared capability in space was taken December 18, 2001 with the declaration of the MCS at Buckley AFB as operationally capable. The MCS consolidates command and control and data processing elements from dispersed legacy systems into a single modern facility. The MCS saves 58% in manning and up to 25% in operations and maintenance costs over the legacy systems. The MCS is also designed to accommodate new SBIRS capabilities.

User Segment

Data from the DSP satellites are processed at the SBIRS MCS and a warning is transmitted to users. The warning information consists of an assessment of the time and place of launch, the type missile launched and an estimated course of the missile. The systems that process DSP data for theater ballistic missile warning are JTAGS and TACDAR.

JTAGS

The Joint Tactical Ground Station (JTAGS) is a transportable information processing system which receives and processes in-theater, raw, wideband infrared data downlinked directly from DSP sensors. The system disseminates warning, alerting, and cueing information on Tactical Ballistic Missiles (TBM), and other tactical events of interest throughout the theater using existing communications networks, primarily IBS broadcast.

Figure 9-2: JTAGS

Developed and built by Aerojet for the US military, JTAGS determines the TBM source by identifying missile launch point and time, and provides an estimation of impact point and time. Since the system is located in-theater, it reduces the possibility of single-point-failure in long-haul communication systems and is responsive to the Theater Commander. It also fulfills the in-theater role of the Air Force Space Command's TES.

JTAGS maximizes the use of commercial off-the-shelf (COTS) and government off-the-shelf (GOTS) equipment. The system is housed in a NBC protected standard military shelter, equipped with a standard wheeled mobilizer that permits tow speeds up to 55 mph by a 5-ton truck. It is air transportable by a C-141 or larger aircraft.

JTAGS can be a key link for the Theater Commander's situational awareness. Operational benefits include:

- Cueing of active theater missile defense systems for missile intercept

- Cueing attack operations assets to find and destroy enemy launch capability

- Timely warning for the protection of friendly forces and population

JTAGS receives direct down linked data from up to three Defense Support Program sensors and follow-on space-based sensors. Features include:

- Threat Tactical Ballistic Missile infrared data

- 3-D stereo processing of multiple sensor downlinks

- Real-time reporting

- Robust multi-networks capability

- In-theater data/voice

The high resolution JTAGS displays include:

- Estimated launch point and time

- Predicted impact point and time

- Trajectory parameter

- Multi-track capability

Space Based Infrared System (SBIRS)

The SBIRS program provides the nation with critical missile defense and warning capability well into the 21st century. SBIRS is one of Air Force Space Command's highest priority space systems. SBIRS will consist of three individual space constellations and an evolving ground element: The Defense Support Program (DSP), SBIRS High, and the Space Tracking and Surveillance System (STSS). These systems are independent yet will complement each other by providing global infrared coverage. The program supports four mission areas: Missile Warning, Missile Defense, Technical Intelligence, and Battlespace Characterization.

SBIRS

SBIRS features a mix of four geosynchronous earth orbit (GEO) satellites, two highly elliptical earth orbit (HEO) payloads, and associated ground hardware and software. SBIRS will have both improved sensor flexibility and sensitivity. Sensors will cover short-wave infrared like its predecessor, expanded mid-wave infrared and see-to-the-ground bands allowing it to perform a broader set of missions as compared to DSP.

The first GEO satellite was launched in May 2011 and has begun delivering infrared imagery to the SBIRS ground station. The satellite includes highly sophisticated scanning and staring sensors that provide wide area surveillance of missile launches and natural phenomena across the globe, while the staring sensor is capable of observing much smaller areas of interest with vastly increased sensitivity. The system tremendously enhances the U.S. military's ability to detect missile launches around the globe, significantly improves technical intelligence gathering capability, and increases situational awareness on the battlefield.

Space Tracking and Surveillance System (STSS)

Originally, SBIRS was to have a low earth orbiting component that was referred to as SBIRS Low. The SBIRS Low program was cancelled and the Space Tracking and Surveillance System was implemented to meet operational requirements. STSS is managed by the Missile Defense Agency (MDA). STSS will build a few satellites at a time with later satellites being more capable than earlier ones. The program will be fully integrated into the nation's ballistic missile defense system architecture, contribute to MDA's ballistic missile testbed, and focus resources on highest leverage technologies. Using the advantage of a lower operational altitude, STSS will track tactical and strategic ballistic missiles. The satellite's sensors will operate across long and short-wave infrared, as well as the visible light spectrum. These wavebands allow the sensors to acquire and track missiles in midcourse as well as during the boost phase, substantially improving the performance of ballistic missile defenses.

STSS is proceeding in a series of biennial "blocks." According to MDA's FY2006 budget documents, STSS' Block 2006 is the launch of what is referred to two legacy satellites; Block 2008 is an improvement of the ground system; and, in Block 2012, operational satellites will be integrated into the program.

Theater Ballistic Missile Warning

Q. What is the threat? Does the enemy have TBM capability?

Q. For what purpose will the command use the warning data received? The timeliness and accuracy required of the warning data provided depends on why the command needs the data and how it will be used. Cueing TMD forces requires quick and accurate data.

Q. How quickly do I need the information at my Operations Center and how quickly do I need to get the information disseminated? This is important because TBM warning information is broadcast to only to certain locations with the necessary receiver equipment. It is the responsibility of the Theater Commander to disseminate the information. Therefore, warning networks must be considered.

Q. What communications networks are available in the AO that can handle voice or data warning reports? Does the command need a separate voice/data system for TBM warning? The proliferation of medium to long-range ballistic missiles and their potential impact on operations may necessitate a separate and more responsive warning network than the normal communications network can provide.

Q. JTAGS needed in theater or can will information from the broadcast networks (TRAP and TIBS) be sufficiently timely and reliable?

Q. Are there areas that would benefit from having terrestrial radar capable of providing theater missile warning? Are resources available to correlate terrestrial radar and satellite warning data? Dual phenomenology provides dual verification and increased accuracy of launch data and refinement of the impact point.

Q. What is the tolerance for false missile warning reports? Missile warning data is gathered from a variety of sources and is then assessed. The results of the assessment can include launch and impact point predictions and a confidence level. The supported commander must consider whether it is more important to have warning of all events (with the possibility of false events) or no false reports with some events not reported.

Q. Will warning reports be provided to allies for broadcast to civilian populations? Dissemination of warning reports to allies is determined by the Theater Combatant Commander or Secretary of State and the Joint Staff in coordination.

Chapter 10
Intelligence, Surveillance, and Reconnaissance (ISR)

INTELLIGENCE, SURVEILLANCE AND RECONNAISSANCE (ISR)

The monitoring of air, land and sea targets from space can provide the military information on enemy locations, dispositions, and intentions. This information can provide warning of enemy attack (to include ballistic missile attack), targeting analysis for deep attack, friendly COA development and Battle Damage Assessment (BDA). This section discusses space support for intelligence, surveillance and reconnaissance, with an emphasis on Imagery (particularly multi-spectral imagery) Intelligence (IMINT) while touching on Signals Intelligence (SIGINT) and Measurement and Signature Intelligence (MASINT). This overall topic may also be referred to as Reconnaissance, Surveillance and Target Acquisition (RISTA). It includes theater ballistic missile detection and warning.

Reconnaissance:

Because space systems have unrestricted overflight of otherwise denied areas, they can gather information about the activities and resources of an enemy or potential enemy.

Surveillance:

Space systems allow commanders to systematically observe and monitor space, surface, or limited subsurface locations or objects at great distances.

OVERVIEW

Analysis of terrain and other environmental factors is a critical step in the Joint Intelligence Preparation of the Operational Environment (JIPOE). The impact of the environment and terrain on the conduct of military operations has been demonstrated throughout history. Without space-based ISR systems, such data can be difficult to obtain, especially in areas where access is limited due to military or political restrictions

Weather and environmental satellites are similar in that they gather information about the nature and condition of the Earth's land, sea and atmosphere by remote sensing. They accomplish this with sensors which observe the Earth in various discrete bands of the electromagnetic spectrum. They are different in that the systems are designed to observe different phenomena and thus have sensors which gather data in different spectral bands with different resolutions. When the data from space systems is merged with that obtained from other ground and airborne sensors, the resultant products are of significantly better quality than those produced from only one source.

With the advent of commercial space-based sub 1-meter resolution imaging capability, there is now a fuzzy distinction between what a remote sensing system is and what a photoreconnaissance system is. Civilian commercial companies in the United States, Russia, India, France, and other countries are selling imagery of various types. Although the commercial companies advertise they are remote sensing programs, the higher resolution images and frequent pass times enables our adversaries access to products that were once only available in the classified national imagery systems. Commercial companies have stated that they will sell as much imagery as is allowed without government intervention. Requesting and buying remote sensing imagery via the Internet is already a growing business. Military commanders and staffs must be aware of what is available to the adversary in peacetime as well as during a crisis.

THE ELECTROMAGNETIC SPECTRUM

The electromagnetic (EM) spectrum is divided into regions based on wavelength from short gamma rays to long radio waves having a wavelength of many kilometers. All objects transmit, absorb or reflect electromagnetic radiation. These characteristics are different for each type of material; therefore each has its own "signature". For example, healthy vegetation has a strong reflectance of infrared light whereas unhealthy vegetation does not. Both, however, may appear to be the same shade of green.

Regions of the spectrum are selected for coverage by sensor bands to optimize collection for certain categories of information most evident in those bands. Multispectral bands in the near infrared (NIR) region and shortwave infrared (SWIR) regions are used to discriminate features that are not visible to the human eye. For instance, actively growing vegetation can be easily separated from many other features in the near infrared (NIR) region because the chlorophyll in the plants is reflected to a far greater extent in

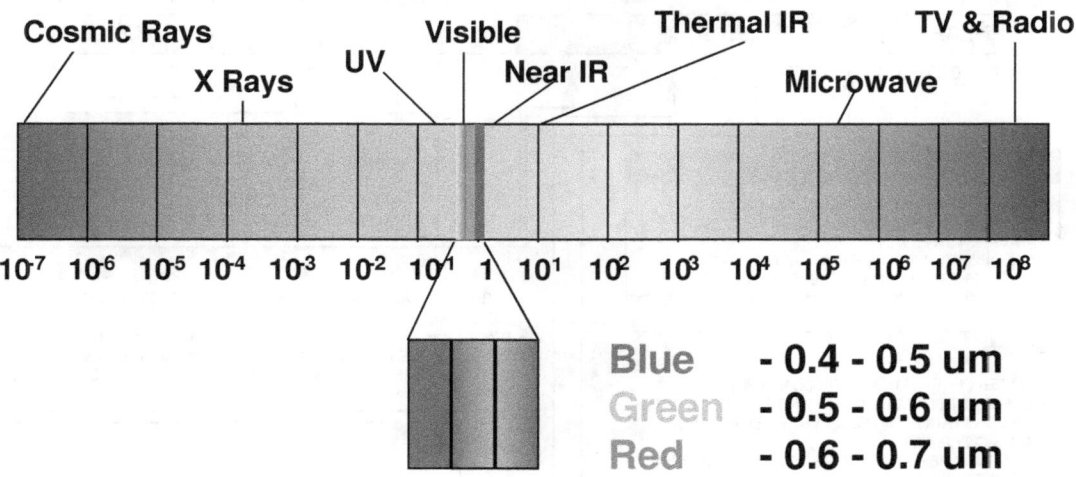

Figure 10-1: Electromagnetic Spectrum

the NIR band than any other feature return. Important features like mineral and oil-bearing rock structures are more easily detected using data collected in these longer wavelength IR bands. Emitted or thermal radiance in the mid-wave and long wave regions is also detectable by passive spectral sensors. For ISR purposes, we know that heat sources such as industrial processes and power generating facilities generate IR energy that are detectable from both aircraft and satellite sensors sensitive to these spectral regions.

The human eye can sense electromagnetic waves with a wavelength between only 0.4 ▢m to 0.7▢m (1 ▢m = 1 micron = 1 x 10^{-6} m = 0.0000001 m). This is visible light. Electronic sensors carried on surveillance and environmental satellites can sense electromagnetic energy across a much greater portion of the spectrum. Each sensor is designed to detect energy in a specific, narrow band of the spectrum.

PANCHROMATIC IMAGERY

Black and white imagery that primarily spans the visual region of the EM (only 2% of the EM spectrum) is called panchromatic imagery. Panchromatic sensors record blue, green and red simultaneously but not separately. Panchromatic imagery displays result in monochrome shades of gray. The human eye can only distinguish approximately 200 shades of gray. Features in panchromatic imagery are interpreted primarily by shape, size, tone, texture, shadow, patter, etc. It is ideal for identifying and analyzing equipment, facilities, and lines of communications. Image manipulation of digital panchromatic imagery can enhance details, sharpen edges, and vary contrast to bring out details which are present, but not evident.

MULTISPECTRAL IMAGERY MSI

Spectral imaging systems are digital - they really aren't pictures at all. The multispectral sensors collect data in digital format. A spectral sensor records the intensity of the energy striking it but only in

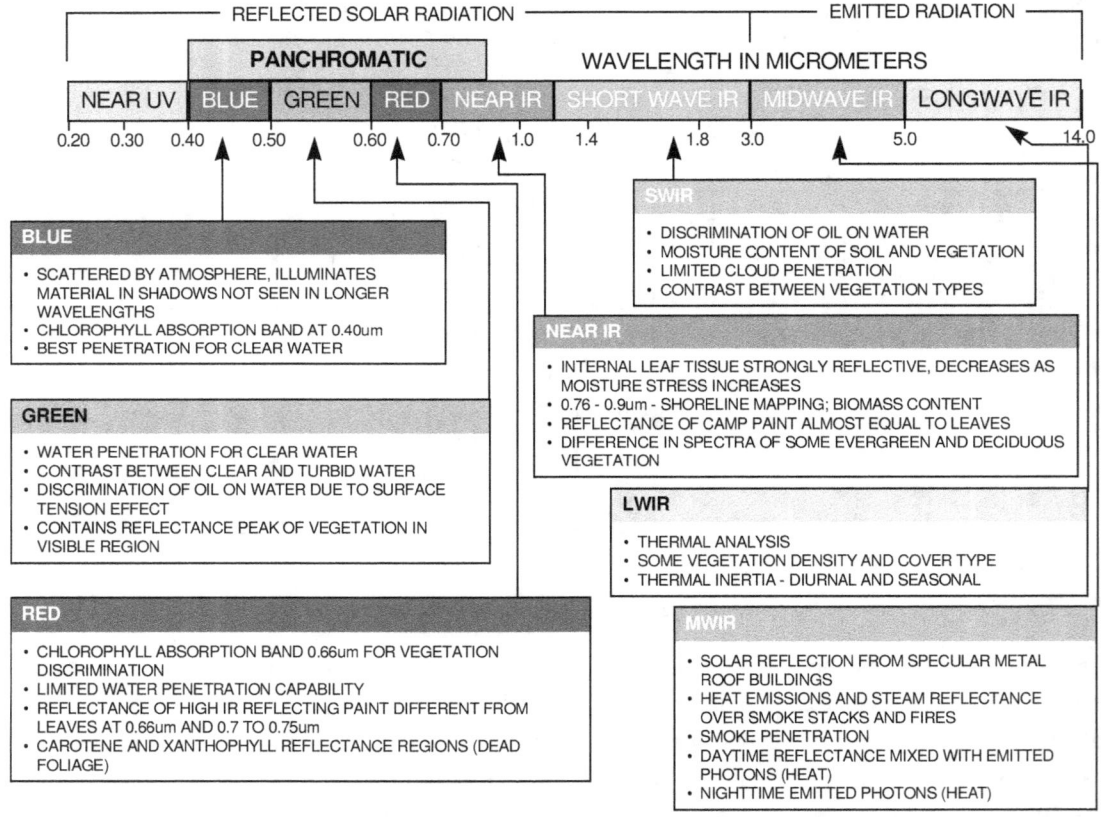

Figure 10-2: Examples of Spectral Regions

the narrow frequency band that it is designed to detect. Two or more spectral sensors are the core of a multispectral sensing system. Energy outside these spectral bands is ignored. Multispectral sensors simultaneously image a scene in numerous electromagnetic bands ranging from visible light through thermal infrared. Multispectral sensors have a much larger dynamic range than panchromatic sensors, thereby producing higher contrast between objects.

The sensors output electronic signals which are converted to digital data. The digital data can be processed in unique ways. For example, haze in an image can be removed by raising the bias level of the signal output. The full dynamic range of the output from the sensors can be digitized and recorded or transmitted to a processing station. After processing, the data from selected sensors are assigned shades of gray or combinations of red, green or blue so that the human eye can see a picture of the data. The resultant image can look significantly different than the normal visible image of the same area. Trained analysts or special computer programs can, however, interpret the image.

Since the information is captured by the sensor as a matrix (rows and columns) of numbers, computers provide a perfect means to sift through the combinations of bands and present the information on a computer monitor. The information can be easily manipulated to support specific analysis. An image can be overlaid electronically with other information, such as slope or land cover data, allowing analysts to derive greater information from the imagery

The number and position of bands in each sensor provide a unique combination of spectral information and are tailored to the requirements the sensor was designed to support. Multispectral sensors are designed to support applications by providing bands that detect information in specific combinations of desirable regions of the spectrum. **Figure 10-2** illustrates the utility of spectral regions. Analysis exploits the spectral separation of reflectance data to detect and identify objects and features.

Each individual object or material has a unique spectral signature based on its reflectance. Spectral imagery records these reflectances and through processing, provides a visual presentation of the reflectance properties. Numeric values (also called brightness values) are recorded as a means of identifying the brightness associated with the light reflected from different material in each spectral band.

Three key points to remember are:

- Data is collected simultaneously in defined bands of the spectrum.

- Each object or material has a unique spectral signature.

- Through processing, the data gathered in the selected bands are processed to discriminate the materials of interest.

Key Terms

There are key terms that must be understood when discussing multispectral imagery. Each term gives different parameters of a multispectral system. Many people do not understand what 30-meter or 1-meter spatial resolution actually means and they equate it to the quality of the image and how much detail that can be seen (i.e. equating it to a National Imagery Interpretability Rating Scale (NIIRS) rating used in reconnaissance and surveillance images).

Resolutions

Spatial Resolution

A spectral collection system records electromagnetic (EM) radiation. The EM radiation detected by the collection system may be solar reflected energy or thermal energy (photons) emitted by an object. EM radiation may be thought of as a wave, having an associated wavelength measured from wave peak to wave peak. A portion of the reflected EM radiation exits through the atmosphere and is received and recorded by the satellite.

This reflected energy is received by the sensor array in the form of individual brightness values for each picture element (pixel). In a digital system, a pixel represents an area on the Earth's surface.

Spatial resolution is another way of stating the size of pixels for a digital system. Pixel size is a direct indicator of the spatial resolution of the sensor because pixels are the smallest elements that can be detected by the sensor. Spatial resolution is a measure of the smallest angular or linear separation between two objects that can be resolved by the sensor. More simply put, it is the smallest separation between two objects on the ground that can be detected as a separate object. This type of resolution is also referred to as the Ground Sampling Distance (GSD) and relates to the size of objects that can be detected on the ground from the sensor. The Satellite Pour l'Observation de la Terre (SPOT) panchromatic sensor has pixels that are the average of the light reflected from a 10 meter by 10 meter area (10m x 10m) on the ground. Therefore, SPOT panchromatic imagery can be said to have 10 meter pixels. An image from the LANDSAT Thermatic Mapper sensor (a US system), which has a GSD of less than 30 meters will not allow for detection of an object that is 5 meters in size. With current systems, resolution is usually referred to in meters and each pixel will sample a square area on the ground in terms of meters.

Spectral Resolution

Spectral resolution refers to the spectral position and bandwidth of a sensor. The LANDSAT TM sensor could be said to have moderate spectral resolution (7 relatively wide bands) in contrast to the SPOT panchromatic sensor which has poor spectral resolution due to its single very broad band. Generally, the narrower the band and the higher the number of bands, the better the spectral resolution will be.

Temporal Resolution

A temporal resolution refers to the time it takes an imaging system to return to an area to collect another image. It is a function of target latitude and satellite orbit. It is essentially a satellite's revisit time to that target. All imagery collected provides an electronic "snapshot" of a particular area and moment. To understand changing conditions of a particular area may require a number of images.

Temporal resolution must be considered when ordering imagery from any source. If a satellite is not over a requester's area of interest when needed, the requested image must wait until that satellite's next pass over the objective area. If the area is cloud-covered during imaging, the user may have to wait for another imaging opportunity. If using LANDSAT 5, the revisit time would be 16 days at the equator. There is also an impact in temporal resolution related to viewing geometry discussed in a following paragraph.

Radiometric Resolution

Radiometric resolution refers to the sensitivity of a spectral band. LANDSAT, SPOT, and Indian Remote Sensing Satellite detect and record an image in each band in 256 levels of brightness (an 8-bit image). One multispectral imager aboard the TIROS weather satellite, called the Advanced High Resolution Radiometer (AVHRR), collects imagery in 1,024 levels of brightness (a 10-bit image). Therefore, TIROS is said to have greater radiometric resolution than LANDSAT, even though the spatial resolution of this sensor is significantly less than that of the others.

Viewing Geometries

Viewing geometries for MSI satellites are available in two varieties: off-nadir/ directional viewing and nadir viewing. Nadir refers to the point on the planet directly below the satellite. Nadir imagers look straight down at the Earth and have no ability to look at objects away from nadir. Such systems are excellent for providing images that have minimal geometric distortions. Distortions in nadir imagers are normally due to Earth curvature and space environmental effects that disrupt the stability of the satellite, introducing pitch, yaw or roll. LANDSAT, being a nadir imager, is restricted to imaging along predictable tracks, or "paths," according to orbital characteristics and the field of view of the sensor.

Directional systems have the ability to view the Earth away from the ground track of the satellite's orbit or, "off-nadir." SPOT is an off-nadir imager with the ability to image up to 27 degrees across track (side-to-side) in each direction. The opportunities presented by off-nadir systems include a reduction in the amount of elapsed time between the periods when the satellite can image the same point on Earth, referred to as "revisit time" or temporal resolution, and the ability to produce stereo views. However, due to the effect of imaging across a spherical surface, users of off-nadir imagery pay the price of higher geometric distortions within the image.

Hyperspectral and Ultraspectral Imagery

An outgrowth of multi-spectral imaging is a cross between imaging and spectroscopy known as imaging spectroscopy or hyperspectral imaging. Hyper and ultraspectral sensors split portions of the electromagnetic range of a sensor into hundreds and thousands of bands.

Hyperspectral Imaging (HSI), like MSI, is a passive technique (i.e., depends upon the sun or some other independent illumination source) but unlike MSI, HSI creates a larger number of images from contiguous, rather than disjoint, regions of the spectrum, typically, with much finer resolution. This increased sampling of the spectrum provides a great increase in information. Many remote sensing tasks which are impractical or impossible with an MSI system can be accomplished with HSI. For example, detection of chemical or biological weapons, BDA of underground structures, and foliage penetration to detect troops and vehicles are just a few potential HSI missions. A multispectral sensor can detect areas covered with grass. A hyperspectral or ultraspectral sensor can distinguish between Kentucky blue grass and Bermuda grass because the differences in the size of the blades of grass and the differences in chemical composition result in slightly different reflectances.

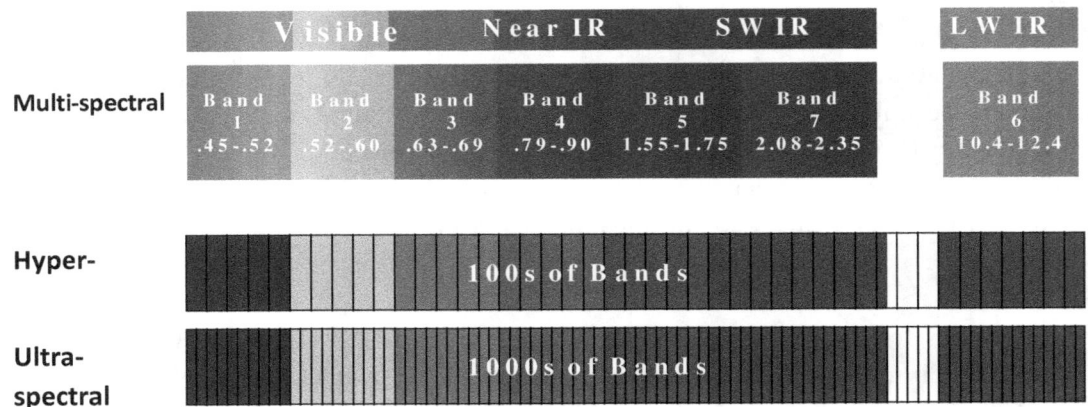

Figure 10-3: Hyper and Ultra-spectral Bands

Hyperspectral remote sensing combines imaging and spectroscopy in a single system which often includes large data sets and requires new processing methods. Hyperspectral data sets are generally composed of about 100 to 200 spectral bands of relatively narrow bandwidths (5-10 nm), whereas, multispectral data sets are usually composed of about 5 to 10 bands of relatively large bandwidths (70-400 nm). This technology is still relatively new. The biggest challenge is the amount of data that must be stored, sent, processed and analyzed. To be able to accurately distinguish subtle differences will require new algorithms and an extensive database of EM signatures of known substances. And even when the technological challenges are resolved, the question of when are 250 bands better than 4 bands will still remain. More bands require more analysis time.

MILITARY IMAGERY APPLICATIONS

Before DESERT SHIELD, MSI was generally treated as a dubious source of terrain information. Institutional DoD production agencies maintained a dogma that MSI was nearly useless as a source of mapping and terrain information. Many service-specific agencies felt the same way. In recent years, an increasing share of training and equipment resources has been dedicated to field users of MSI, such as USMC and US Army topographic units. USAF and US Navy applications of MSI have also been gaining acceptance, supporting image mapping, mission planning, navigation and targeting. Historically, these products and others have been widely used in operations such as DESERT SHIELD/DESERT STORM, support to activities in the former Yugoslavia, Somalia, Haiti, many "special" operations and in a large number of training exercises. With the ensuing Operations Iraqi Freedom and Enduring Freedom, MSI became quite useful among tactical forces and other agencies that soon provided MSI products. MSI developed into a useful tool and provided map supplements to out-of-date maps as well as a terrain database.

Although not a complete list, the following MSI applications are among the more commonly used by today's military.

Analysis Images are among the simplest of MSI products, normally consisting of a natural colored image of the desired site. The image is minimally processed to reduce the production time and is constructed in pre-determined sizes according to need. Camouflage that couldn't be detected in panchromatic images, such as netting, can be detected using the NIR and SWIR bands. One can also detect vegetation stress and find concealed objects.

Context Images are similar to Analysis Images but are frequently produced in black and white to facilitate correlation with black and white high resolution images used by intelligence analysts. The purpose of Context Images is to provide a broad area perspective around a target or the AOR with a picture that is larger than most near-real-time systems.

Image Map or "I-Map" is a common application of Multispectral Imagery and exploit the broad-area coverage capabilities of MSI. An I-Map is nothing more than an image of the area of interest with a commonly understood grid overlaid. Typically, the image is rectified so that features on the Image Map correspond to features on a selected coordinate system and are in proper relationship to features on the Earth's surface. The advantage of this product is that the user receives a literal image that requires little experience or training to use. This type of image portrays the terrain in near-natural colors, with some enhancements applied to ease use by non-trained personnel. An image map is most useful when a topographic map is not available or when used as a supplement to an older map. Multispectral Imagery provides a means of rapidly updating an older, out-of-date, topographic map with an image map that provides a current, broad-area, synoptic view of the area of operation. The map is processed to user specifications in terms of color, size, map projections and scale.

Image Mosaics are produced from two or more products and made into one larger geographic area. Mosaics are produced by digitally "stitching" images together, usually for visual effect. Prior to "stitching," images are pre-processed to ensure an acceptable match of brightness value ranges (histogram matching) and that bands selected for the final product appear visually similar.

Perspective Views are a combination of Multispectral Imagery and elevation data such as Digital Terrain Elevation Data (DTED) and Digital Elevation Model (DEM) (produced and maintained under the auspices of the NGA). The two-dimensional MSI image is draped over the three-dimensional digital terrain data and processed to simulate a view of an area or target of interest from a given position, altitude, azimuth and distance. This can be used to produce 3D perspectives and animated fly-throughs for mission planning or rehearsals.

Relocatable Target Graphics are developed from many sources of information including MSI. In this product, several inputs are combined and evaluated using geographic information system techniques to enable prediction of the movements of mobile targets. Elevation matrices are converted to slope and evaluated to determine places where mobility restrictions would prevent target movement. MSI data is thematically classified to further reduce the potential hiding areas. Known information about operating characteristics of relocatable mobile targets is included in the analysis to provide predictive movement information. High resolution imagery is used to evaluate areas too small to be identified through other processes.

The data is combined in a processing scheme to predict the movement of vehicles given a known point of origin and terrain factors revealed through the analysis of the above data. Graphics registered to a selected map base are then produced to indicate the probability of movement across the terrain. Both hardcopy and softcopy products are constructed in this manner, depending upon user requirements.

Stereo Imagery is used to create elevation matrices and is made possible by off-nadir imaging systems or, less frequently, by two images from a single nadir imaging system. Two images of the same ground area are captured from different points in space (producing a stereo view of the area) and software is used to establish a mathematical relationship between points that can be identified on each image. Elevation data is then extracted using specially designed software and placed in a file to be used as needed.

TERCAT (TERrain CATegorization) is a pseudo-color thematic image in which the Multispectral Imagery data has been classified into groups representing different terrain types and land cover. TERCATs are useful in displaying vegetation classes, soil types and hydrology that affect trafficability. They also support the analysis of lines of communications (LOCs), avenues of approach, cover and concealment, landing and drop zones.

Terrain Analysis Products are emerging as an important MSI product, drawing upon the ability of MSI to "thematically classify" statistically similar brightness values representing various types of terrain occurring in the image. Land-cover types (vegetation, urban areas, water) and density information are locked in the spectral images awaiting extraction using a variety of automated techniques and human judgment. Although terrain analysis products derived from MSI are not as accurate as "objective terrain databases," in most cases, they can be developed much more quickly and within acceptable tolerances for supporting many tactical and strategic activities.

Hydrology and Bathymetry – depths of near-shore channels, reefs, and underwater obstacles can be determined down to about 50 feet in clear water. However, turbid water and submerged vegetation can degrade the accuracy of water depths.

Change detection involves using multispectral imagery over time to record changes in the environment.

REMOTE SENSING AND SATELLITE SYSTEMS

Remote Sensing

Remote sensing and the MSI sensor offer other advantages than just being able to image in specific spectral regions. These include the ability to "capture" images digitally, transmit the imagery electronically, store the information on magnetic media and process the digital information using computers.

Remote sensing gives a unique view of the Earth. Remote sensing by satellites, especially multispectral imagery satellites, provides critical information that is of immediate military value over large areas. Remote sensing also has numerous civil and commercial applications which have made it one of the fastest growing areas of space after communications.

Remote sensing is not limited to only satellites nor is it limited to multispectral systems. Remote sensing can be done from virtually any airborne platform. Additional remote sensing capabilities include specific infra-red (IR) sensors and synthetic aperture radar (SAR) capability. SAR uses radar technology in a very specific way. To understand how it works, it is necessary to first understand the basics of radar.

Typical radar (RAdio Detection and Ranging) measures the strength and round-trip time of the microwave signals that are emitted by a radar antenna and reflected off a distant surface or object. The radar antenna alternately transmits and receives pulses at particular microwave wavelengths (in the range 1 cm to 1 m, which corresponds to a frequency range of about 300 MHz to 30 GHz) and polarizations (waves polarized in a single vertical or horizontal plane). For an imaging radar system, about 1500 high-power pulses per second are transmitted toward the target or imaging area, with each pulse having a pulse duration (pulse width) of typically 10-50 microseconds (us). The pulse normally covers a small band of frequencies, centered on the frequency selected for the radar. At the Earth's surface, the energy in the radar pulse is scattered in all directions, with some reflected back toward the antenna. This backscatter returns to the radar as a weaker radar echo and is received by the antenna in a specific polarization (horizontal or vertical, not necessarily the same as the transmitted pulse). Given that the radar pulse travels at the speed of light, it is relatively straightforward to use the measured time for the roundtrip of a particular pulse to calculate the distance or range to the reflecting object. The chosen pulse bandwidth determines the resolution in the range (cross-track) direction. Higher bandwidth means finer resolution in this dimension. The length of the radar antenna determines the resolution in the azimuth (along-track) direction of the image: the longer the antenna, the finer the resolution in this dimension.

Synthetic Aperture Radar (SAR) refers to a technique used to synthesize a very long antenna by combining signals (echoes) received by the radar as it moves along its flight track. Aperture means the opening used to collect the reflected energy that is used to form an image. In the case of a camera, this would be the shutter opening; for radar it is the antenna. A synthetic aperture is constructed by moving a real aperture or antenna through a series of positions along the flight or orbital track of the host vehicle.

Satellite Systems

There are over 15 remote sensing/earth observation satellites owned by the US, Russia, France, India, Brazil, China, the European Union, Israel, Japan, and Korea. Many of these countries plan to compete directly with American civil and commercial systems.

The discussion that follows should not be considered all-inclusive.

LANDSAT (US)

The first satellite to use multispectral imagery for Earth remote sensing was the Earth Resources Technology Satellite-A (ERTS-A) which was later renamed Landsat-1. It was the first in a series of satellites designed to provide repetitive global coverage of the Earth's land masses. ERTS-A was launched in July 1972 and finally ceased to operate in January 1978. NASA was responsible for operating the LANDSAT satellites through the early 1980's. In 1985, as part of an effort to precipitate commercialization of space, LANDSAT was commercialized. On

Figure 10-4: LANDSAT

April 15, 1999 the latest member of the Landsat family, Landsat 7, was launched into orbit.

The current constellation consists of two satellites, Landsats 5 and 7. They are in sun-synchronous orbits at 438 miles (705 km) altitude and a sun-synchronous 98.2 inclination with ground tracks of 185 km width. This allows a satellite to pass within sensor range of every location on Earth at least once every 16 days. The Landsat system provides for global data between 81 degrees north latitude and 81 degrees south latitude. This final maneuvering process will place Landsat 7 in an orbit consistent with and eight paths offset to the east of Landsat 5 paths. This will result in a Landsat 7 and Landsat 5 overflight of the same location eight days apart during routine operations.

CHARACTERISTICS OF THEMATIC MAPPER BANDS		
Band	Wavelength, μm	Characteristics
1	0.45 - 0.52	Senses blue-green visible light. Maximum penetration of water which is useful for mapping in shallow water. Also useful for distinguishing soil from vegetation and deciduous from coniferous plants.
2	0.52 - 0.60	Senses green visible light. Matches green reflectance peak of vegetation. Useful in assessing plant vigor.
3	0.63 - 0.69	Senses red visible light. Matches chlorophyll absorption band. Useful in discriminating vegetation types.
4	0.76 - 0.90	Senses reflected near infrared. Useful for determining biomass content and for mapping of bodies of water which appear opaque.
5	1.55 - 1.75	Senses reflected mid-infrared. Indicates moisture content of soil and vegetation. Penetrates thin clouds. Good contrast between vegetation types. Useful for differentiation between snow and clouds.
6	10.40 - 12.50	Senses thermal infrared. Can be used at night. Useful for thermal mapping and estimating soil moisture.
7	2.08 - 2.35	Senses reflected infrared. Wavelength coincides with an absorption band of hydroxyl ions in minerals. Combination of bands 5 and 7 useful for mapping hydrothermally altered rocks associated with mineral deposits.

Table 10-1: Thematic Mapper Band Characteristics

Each Landsat satellite carries the Multispectral Sensor (MSS), the Thematic Mapper (TM), and now the Enhanced Thematic Mapper Plus (EMT+). The TM sensor, available on Landsat 5, captures information in seven spectral bands. The EMT+, available on Landsat 7, adds a panchromatic band. The MSS and TM sensors primarily detect reflected radiation from the Earth's surface in the visible and IR wavelengths, but the TM sensor provides more radiometric information than the MSS sensor.

The Multispectral Scanner collects data by continuously scanning Earth from west to east using an oscillating mirror, recording radiation in four different spectral bands in the visible and the near-IR regions with a resolution of 80 m. Bands 1 and 2 are in the visible wavelengths; bands 3 and 4 are in the near-IR portion of the spectrum. In order to present all of the bands in a fashion visible to the human eye, colors or shades of black and white are assigned to each band with the result of creating an image. As an example: band 1 detects green and could be shown as blue; band 2 detects red and could be shown as green; either band 3 or band 4, both of which detect separate bands of IR light, can be used in the construction of an image with bands 1 and 2 and could be shown as red.

The Thematic Mapper receives solar reflected energy covering the entire visible spectrum and the near, shortwave, and thermal-IR spectrums using 7 bands. In this respect, the Thematic Mapper sensor has the greatest spectral discrimination of any sensor platform currently in orbit. This capability allows greater discrimination among a large number of terrain feature types. Six of these bands provide at least 20 basic combinations and 120 permutations for a three-band color image. Each of these combinations has its own unique attributes. An optimum combination is a function of the intended application and a matter of personal band combination for imagery analysis. The TM sensor has a spatial resolution of 30 meters for the visible, near-IR, and mid-IR wavelengths and a spatial resolution of 120 meters for the thermal-IR band.

There are a few differences between Landsat 5 and Landsat 7. One of the Landsat 7 improvements is the Enhanced Thematic Mapper Plus (ETM+). The ETM+ instrument is an eight-band multispectral scanning radiometer capable of providing high-resolution imaging information of the Earth's surface. It detects spectrally-filtered radiation at visible, near-infrared, short-wave, and thermal infrared frequency bands from the sun-lit Earth. Band 8 is a panchromatic band with a resolution of 15 meters. Band 6 now has both high gain and low gain settings at 60 meter resolution. There are 3 on-board solar calibrators to provide an absolute radiometric accuracy of +/-5%. In addition there is a "cloud cover predict" implemented in the Long Term Acquisition Plan, and a more accurate cloud cover assessment performed on acquired data. Nominal ground sample distances or "pixel" sizes are 49 feet (15 meters) in the panchromatic band; 98 feet (30 meters) in the 6 visible, near and short-wave infrared bands, and 197 feet (60 meters) in the thermal infrared band.

On 31 May 2003 Landsat 7's ETM+ sensor's Scan Line Corrector suffered a failure. Without the SLC, some scan lines in the pictures taken by the ETM Plus overlap while others are missing. In total, about 30 percent of data in each picture is lost. Most of the missing regions in the pictures can be filled by interpolation; however, this is an approximation rather than real data.

As a commercial system, only licensed ground stations can receive and process Landsat data. The primary ground station, the data handling facility and the archive are located at the USGS/EROS Data Center in Sioux Falls, SD. The ground system can distribute raw ETM+ data within 24 hours of its reception at the EROS Data Center. Secondary USG ground stations for Landsat 7 are located in the Fairbanks, Alaska area and Svalbard, Norway. The secondary station(s) serve as backup for the primary station and ensure that the requirement for scene acquisitions is met. In addition, the Landsat 7 system is capable of transmitting data in real-time to other, non-USG ground stations. These stations will receive direct downlink data upon request.

Systeme Probatoire d'Observation de la Terre (SPOT) (France)

SPOT satellites are owned and operated by the Centre National d'Etudes Spatiales (CNES), a French governmental organization.

The satellites have the same mission as Landsat but have different capabilities. They are launched into a sun synchronous 822 km high orbit with an inclination of 98.7°. The satellite repeats its ground trace every 26 days. In the panchromatic mode (black and white) visual images with a resolution of 10 meters are possible. The four band multispectral mode has a resolution of 20 meters.

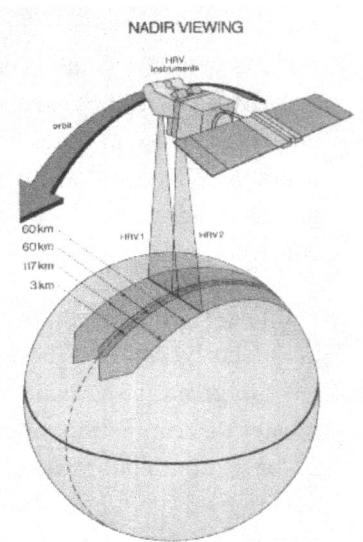

SPOT 1 was launched in 1986. SPOT 1 is still operational but has been placed in an inactive role.

The SPOT 2 satellite was launched on 22 January 1990, it was the first in the series to carry the DORIS precision positioning instrument.

Figure 10-5: SPOT

SPOT 3 followed on 26 September 1993. It also carried the DORIS instrument, plus the American passenger payload POAM II, used to measure atmospheric ozone at the poles. SPOT 3 failed in 1996, having reached the end of its nominal mission lifetime.

SPOT-4 was launched in March 1998 with a designed for a five year life span and is still operational. It carries two identical high resolution sensors, called HRVIR (High Resolution Visible – Infrared), which operate in the visible and near infrared ranges of the electromagnetic spectrum. Each sensor contains a static solid-state linear array of detectors that work together as a "push-broom" scanner. As the spacecraft moves forward, each sensor scans side-to-side by means of a steerable mirror. HRVIR includes a new medium IR channel to support vegetation analysis and harvest forecasting. The Vegetation Monitoring instrument has 1 km resolution in the same bands as the HRVIR.

SPOT 5 was launched in May 2002. The SPOT 5 satellite offers improved ground resolutions of 10 meters in multispectral mode and 2.5 to 5 meters in panchromatic and infrared mode. Higher 2.5-meter resolution is achieved using an innovative sampling concept called Supermode.

SPOT 5 also features a new HRS imaging instrument (High Resolution Stereoscopic) operating in panchromatic mode and able to point forward and aft of the satellite. In a single pass, the forward-pointing camera acquires images of the ground, and then the rearward-pointing camera covers the same strip 90 seconds later. HRS is thus able to acquire stereo pair images almost simultaneously to map relief, produce digital elevation models (DEMs) of wide areas and generate orthorectified products of a quality unequalled on the market today. The HRS instrument can acquire up to 126,000 square kilometers of data every day, thus providing a fast turnaround for users.

SPOT 5 also carries a VEGETATION 2 instrument identical to that on SPOT 4.

The HRVIRs are steerable to within 27 deg off-nadir which means they can point to spots on the Earth's surface between +27° and -27 ° from the vertical (extending from the satellite to its ground track) as the satellite moves forward. The maximum ground swath (swath width) for one sensor operating alone is 60 km. The overlap, over the satellite's ground track, of the two swaths is 3 km, giving a total ground swath of 117 km when the two sensors operate at the same time.

Because SPOT has oblique viewing capabilities, stereoscopic imaging is a major feature of the satellite. Stereoscopic pairs of images are commonly used in topographic mapping and geomorphologic and geological studies because they provide high resolution in nearly 3-D images. Two images of the same area acquired on different dates and at different HRV angles can be combined to produce a vertical view that gives very accurate topographic information (i.e., altitude). This capability also allows for viewing of the same area every three or four days although from different angles.

When compared to Landsat, the spatial resolution of SPOT images is higher than that of Landsat; however, Landsat images have more spectral bands. SPOT sensors can also provide stereo image data. Both sets of data can be purchased in the commercial market. During Operation DESERT STORM, France implemented restrictions on the distribution of SPOT data to members or supporters of the coalition of allied forces, of which France was a member. The Army integrated SPOT and Landsat data which resulted in a significantly better product. The SPOT data most frequently used by the DoD is the HRS panchromatic imagery.

The SPOT Mission Center is located in Toulouse, France. A second command center is located at Kiruna, Sweden. At these sites ground operators collect real-time images, within 2500km of the ground stations, and stored images from other parts of the Earth. The two ground stations have the capacity of receiving 500,000 images per year. SPOT Image, the company which runs the SPOT marketing, runs a number of Direct Receiving Stations which receive real-time images only. These stations are located in Canada (2), India, Spain, Brazil, Thailand, Japan, Pakistan, Saudi Arabia, South Africa, Australia, Ecuador, Taiwan, Indonesia, and Israel. Data transmitted in direct mode are received in real time during daytime passes. Data transmitted in recording mode (i.e., via on-board tape recorders) are received during nighttime passes by SPOT's main receiving stations in Toulouse, France, and Kiruna, Sweden. SPOT imagery is sold through Spot Image Corporation in Reston, VA. There are also SPOT commercial sales offices in France, Australia, Singapore, and China.

Indian Remote Sensing (IRS) Satellite (India)

The launching of the first Indian Remote Sensing Satellite, IRS-1A in March 1988, followed by the successful launch, calibration, and initialization of IRS-1B in August 1991, IRS-P2 during 1994, IRS-1C during 1995, IRS-P3 during 1996, IRS-1D during 1997, and IRS-P4 during 1999, has provided India with an unique opportunity to use remote sensing data for the monitoring and management of natural resources and the environment. India sells this data to customers around the world.

IRS-1D is equipped with three sensors that collect five-meter resolution panchromatic (black-and-white), 20-meter resolution multispectral (color), and 180-meter resolution multispectral wide-area images. The IRS-1D panchromatic sensor sacrifices swath width for its higher resolution; however, it can be pointed off the orbit path which allows 2 to 4 day revisits to specific sites. There are on-board recorders for data collection outside the range of ground station, to further increase data availability.

IRS-P4 (OCEANSAT-1) has a Multi-frequency Scanning Microwave Radiometre (MSMR) operating in four frequencies and a nine-band Ocean Colour Monitor (OCM). It was launched in May 1999. The satellite's sensors have a resolution of 250 meters at nadir and have a swath width of 1420 km with a revisit time of two days. The payload is specifically tailored for the measurements of physical and biological oceanography parameters. It provides valuable Ocean-Surface related observation capability. It has a multispectral spatial resolution of 32 meters and panchromatic spatial resolution of 5 meter. The sensors can steer along the ground track at +/- 20 degrees in steps of 5 degrees.

The IRS-P5 (CARTOSAT-1), launched in May 2005, carries two state-of-the-art panchromatic (PAN) cameras that take black and white stereoscopic pictures of the earth in the visible region of the electromagnetic spectrum. The swath covered by these high resolution PAN cameras is 30 km and their spatial resolution is 2.5 meters. The cameras are mounted on the satellite in such a way that near simultaneous imaging of the same area from two different angles is possible. This facilitates the generation of accurate three-dimensional maps. The cameras are steerable across the direction of the satellite's movement to facilitate the imaging of an area more frequently. The images taken by CARTOSAT-1 cameras are compressed, encrypted, formatted and transmitted to the ground stations. The images are reconstructed from the data received at the ground stations. It also carries a solid state recorder with a capacity of 120 Giga Bits to store the images taken by its cameras. The stored images can be transmitted when the satellite comes within the visibility zone of a ground station.

IRS-P7 (CARTOSAT-2) carries a state-of-the-art panchromatic (PAN) camera that take black and white pictures of the Earth in the visible region of the electromagnetic spectrum. The swath covered by the high resolution PAN camera is 9.6 km and the spatial resolution is less than 1 meter. The satellite can be steered up to 45 degrees along as well as across the track. It can produce images of up to 80 cm in resolution, compared to the 100 cm (1 m) offered by Ikonos. In the past, India has been buying images from Ikonos (described later) at about $20 per square meter of imagery. With Cartosat-2 offering better resolution at twenty times lower cost per sq m of imagery, buying images from Ikonos is likely to decline in future.

There is a business alliance between the Indian ANTRIX Corporation which is a commercial organization and the Lockheed Martin/Space Imaging-EOSAT. Under this alliance, some of the global stations in the US, Germany, Thailand, etc. are receiving IRS data for distribution to the global customers.

The National Remote Sensing Agency (NRSA) is an autonomous organization under Department of Space, Govt. of India engaged in operational remote sensing activities. NRSA has its own ground station at Shadnagar, 60 km south of Hyderabad to acquire remote sensing satellite data from Indian Remote Sensing satellites, the latest being IRS-P4, and other foreign satellites like Landsat, NOAA, and ERS.

RADARSAT (Canada)

RADARSAT is a cooperative program between the Canadian Space Agency, NASA and NOAA. The Canadian Space Agency built and operated the satellite; NASA furnished the launch. In exchange, US government agencies will have access to all archived RADARSAT data and have approximately 15% of the satellite's observing time. RADARSAT-1 was launched in 1995 and is equipped with an advanced

Synthetic Aperture Radar (SAR) instrument capable of delivering data from 7 beam modes and 25 beam positions. RADARSAT-2 was launched in December 2007 with first images available in January 2008.

A SAR is a powerful microwave instrument that transmits pulsed signals to Earth and then processes the returned signals. SAR-based technology provides its own illumination, enabling it to penetrate through clouds, haze, smoke and darkness, thus providing images of the Earth in all weather conditions, at any time. This ability offers a much needed alternative during periods when cloud cover prevents imaging with a passive sensor (sensors that do not provide their own illumination) such as LANDSAT, SPOT and other ISR satellites.

RADARSAT is in a sun-synchronous polar orbit at 98.6 degrees. It provides the first routine surveillance of the entire Arctic region and accurately monitors disasters such as oil spills, floods and earthquakes. RADARSAT provides radar imagery which is of exceptional value in supporting a wide range of DoD and civilian applications. Fine resolution is 8m, standard resolution is 30m and ScanSAR wide is 100m. The satellite pass is every 24 days but it can provide daily coverage of the artic region and any part of Canada within 3 days. Data can be processed and delivered in less than 4 hours after it has been acquired.

RADARSAT International, Inc. will be the commercial distributor of RADARSAT data worldwide. Lockheed Martin has distribution rights in the United States.

RESURS F and DK (Russia)

The former Soviet Union launched numerous Resurs-F remote sensing satellites. Russia has continued the program but with infrequent launches. The F2-series satellites has about a 1 month lifetime while the F1 series has only half that. The results of the mission (films) are sent back to Earth in a reentry vehicle. This remote sensing satellite program is operated by the Priroda ('Nature') center of the Russian Central Geodesy and Cartography Agency (GUGK).

The satellites are launched into a low earth polar orbit. The satellites are used for civil remote sensing photography. Resolution of these images is about 5 meters. The apparent principal use is to monitor the condition of agricultural areas within Russia and other former Soviet republics. The Russians also sell some photographs on the commercial market if the images are not of areas they consider sensitive. Although the resolution is high, the data is difficult to integrate with digital data from other multispectral satellites since the only products are photographs (a form of analog data).

The latest in the series is the RESURS DK vehicle. Information about the series is not widely available, but launching began in the 1990s. There launch of a RESURS DK-1 was in June 2006.

RESURS O1

The first satellite in the RESURS-O1 series was operational for three years after its launch in 1985. The second, launched in 1988, was switched off after 7 years of successful operation. The third satellite was launched in 1994. It is still partly operational, but it does not receive any transmissions on X-band anymore. The successor was launched in July 1998, but both of the transmitters on-board have failed and we cannot receive any data at all from it.

RESURS-O1 #3 has two remote sensing instruments on board. The MSU-E is comparable with instruments on other satellites, such as Landsat, while the MSU-SK offers perspectives of the Earth that have never been available before. The MSU-E offers three spectral bands and spatial resolution of 45 meters. The MSU-SK is a five spectral band (four visible and one NIR) instrument offering160 meter resolution in the multispectral range and 600 meter resolution in the NIR range.

Through agreements between Satellus and its Distributors, digital Quick Looks and other information are available about the archived RESURS-O1 images.

Other Imagery Systems

China-Brazil Earth Resources Satellite (CBERS or Zi Yuan 1) (China and Brazil)

China and Brazil agreed on July 6, 1988, to start a cooperative program to develop two remote sensing satellites. The first satellite was launched in October 1999 and the second in 2003. The Wide Field Imager has a ground swath of 890 km which provides a synoptic view with spatial resolution of 260 m. The Earth surface is completely covered in about 5 days in two spectral bands: 0.66 mm (green) and 0.83 mm (near infra-red). The high-resolution CCD camera provides images of a 113 km wide strip with 20 m spatial resolution. Since this camera has a sideways pointing capability of ±32 degrees it is capable of taking stereoscopic images of a certain region. The CCD camera operates in 5 spectral bands that include a panchromatic band from 0.51 to 0.73 mm. The two spectral bands of the WFI are also present in the CCD camera to allow complementing the data of the two types of remote sensing images. A complete coverage cycle of the CCD camera takes 26 days. Infrared Multispectral Scanner (IR-MSS) operates in 4 spectral bands such as to extend the CBERS spectral coverage up to the thermal infrared range. It images a 120 km swath with a resolution of 80 m (160 m in the thermal channel). In 26 days one obtains a complete Earth coverage that can be correlated with the images of the CCD camera.

This joint venture has proven very successful. With already 3 satellites launched, the latest being CBERS-2B launched in September 2007, there are currently plans for CBERS-3 and CBERS-4 in the future.

Earth Remote Sensing Satellite (ERS) (Europe)

The European Space Agency, a consortium of European countries, owns and operates the ERS satellites. The last ERS (ERS-2) satellite was launched in 1995. It is designed to produce images and other data on the ocean surface, ocean temperature, ocean bottom in shallow coastal areas, wave patterns, ice conditions, crop development and forest management. The Synthetic Aperture Radar (SAR) instrument is designed primarily to measure ocean wave length and direction but has also been able to detect ocean wave fronts and current shear. ERS-2 operates in the IR and visible spectrum.

ERS-2 is a notable European engineering achievement, reaching the milestone of 10 years in orbit on 21 April 2005 with all instruments still working and providing excellent data. Over this ten year period the satellite has underpinned and supported the development of a unique know-how, a broad range of outstanding Earth observation science results and a range of operational applications.

The European Space Agency followed the ERS mission with ENVISAT in March 2002. The ENVISAT satellite has an ambitious and innovative payload that will ensure the continuity of the data measurements of the ESA ERS satellites. ENVISAT data supports Earth science research and allows monitoring of the evolution of environmental and climatic changes. The data facilitate the development of operational and commercial applications.

Helios (France, Italy, Spain)

Helios is a French military observation program built in concert with Italy and Spain. The Helios-1 payload consists of high resolution optical observation equipment capable of 1m resolution and two tape recorders. It was launched in 1995. Helios 1B was launched in Dec 1999. The Helios-1 satellites cannot observe at night or through clouds. Each country can use the satellite in proportion of its financial participation. Each country can order observations and obtain the results via its Earth station. Each country's results are encrypted so that only the orderer can use the data. However, France may find itself compelled to put images from its Helios-1, and possibly Helios-2, satellite on the commercial marketplace to counter a US move to sell commercial images with a resolution of one meter. Both satellites are still operating.

GeoEye (US)

IKONOS

The IKONOS satellite is the world's first commercial satellite to collect black-and-white (panchromatic) images with 1-meter resolution, and multispectral imagery with 4-meter resolution. Imagery from the panchromatic and multispectral sensors can be merged to create 1-meter color imagery (pan-sharpened). To date, IKONOS has collected more than 250 million square kilometers of imagery over every continent. IKONOS imagery is being used for national security, military mapping, air and marine transportation, and by regional and local governments. The IKONOS satellite weighs about 1600 pounds. From a 423 mile (680 km) high, sun-synchronous orbit, IKONOS passes any given longitude at about the same local time (10:30 A.M.) daily and has a revisit time of once every three days. The satellite downlinks directly to more than a dozen ground stations around the globe.

ORBVIEW-2

Launched in 1997, OrbView-2 collects color imagery of the Earth's entire land and ocean surfaces on a daily basis. As the foundation upon which GeoEye built the SeaStar Fisheries Information Service, OrbView-2 saves time, fuel and money for commercial fishing vessels by helping them find fish faster. OrbView-2 detects changing oceanographic conditions used to create fishing maps that are delivered directly to commercial fishing captains at sea. The satellite also provides broad-area coverage in 2,800 km-wide swaths, which are routinely used in naval operations, environmental monitoring, and global crop assessment applications.

GeoEye-1

GeoEye will continue its tradition of mapping, monitoring, and measuring the Earth's surface with the launch of GeoEye-1, its next-generation satellite with the highest resolution and most advanced collection capabilities of any commercial imaging satellite. Scheduled for launch in 2008, GeoEye-1 will offer unprecedented spatial resolution by simultaneously acquiring 0.41-meter panchromatic and 1.64-meter multispectral imagery. It can collect 700,000 square kilometers of imagery in a single day, downlink imagery in real-time to international ground station customers, and can store 1.0 terabytes of data on its solid-state recorders. The detail and geospatial accuracy of GeoEye-1 imagery will further expand the applications for satellite imagery in every commercial and government market sector.

KITSAT (South Korea)

KITSAT 3, developed in a university research center, is a micro-satellite that was launched in May 1999 on the Indian PSLV. Previous KITSAT satellites did not include multispectral imaging systems. The current satellite has a remote sensing system capable of acquiring images in three bands – red, green and NIR. Spatial resolution is 15m with a swath width of 50km.

KOMETA (Russia)

Although not a multispectral system, the KOMETA is a space-based mapping system. It is composed of the topographic camera TK-350 and the high resolution camera KVR-1000, integrated with on-board equipment for external orientation elements determination, is designed to provide large scale topographic and digital maps. The satellite operates in near-circular orbit at an average altitude of 220 kilometers and has an orbital duration of 45 days. The entire system is retrieved from orbit and landed at the preliminary defined location. The film is then retrieved from the descending module, processed, converted into digital form using precise scanners, and orthorectified. The digital products are delivered to customers on CD-ROM.

The SPIN-2 is a trademark for Russian digital orthorectified geo-coded two meter resolution satellite imagery data. SOVINFORMSPUTNIK, Aerial Images, Inc., and Central Trading Systems, Inc. have contracted to jointly market high resolution panchromatic Russian satellite imagery data. This unique project brings to the civilian market, for the first time, cartographic quality, high resolution image data collected from formerly classified Russian military satellite systems.

Okean (Russia and Ukraine)

The Okean-O satellite was launched in July 1999. The Okean program has been in existence since 1983. The satellites now provide multispectral, radar, microwave and optical images of Earth for the purposes of sea navigation, fishery and coastal shelf usage. The Okean-O has 2 side-looking radars, 2 tracking UHF radiometers. It is capable of collecting information in eight spectral bands with resolutions down to 25m depending on the imager used.

Terra (EOS-AM1)

Terra (formerly known as EOS-AM1) is a NASA scientific research satellite launched in December 1999. The instrumentation package includes a MODIS Moderate Resolution Spectro-radiometer (MRS) which is capable of imaging in 36 spectral bands from visible to thermal infrared. It can view the entire earth's surface every 1 to 2 days.

A second multispectral instrument on board is the Multi-angle Imaging Spectro-Radiometer (MISR). MISR is a new type of instrument designed to view the Earth with cameras pointed at nine different angles. One camera points toward nadir, and the others provide forward and aftward view angles, at the Earth's surface, of 26.1°, 45.6°, 60.0°, and 70.5°. As the instrument flies overhead, each region of the Earth's surface is successively imaged by all nine cameras in each of four wavelengths (blue, green, red, and near-infrared). MISR data can distinguish different types of clouds, aerosol particles, and surfaces.

The third multispectral instrument is the Advanced Spaceborne Thermal Emission and Reflection Radiometer (ASTER). ASTER will obtain high-resolution (15 to 90 m) images of the Earth in the visible, near-infrared (VNIR), shortwave-infrared (SWIR), and thermal-infrared (TIR) regions of the spectrum. ASTER consists of three distinct telescope subsystems: VNIR, SWIR, and TIR. It is a high spatial, spectral, and radiometric resolution, 14-band imaging radiometer. All three ASTER telescopes (VNIR, SWIR, and TIR) are pointable in the crosstrack direction and the telescopes do not have a zoom capability. The pixel resolution is fixed within each telescope, but varies between the telescopes. Given its high resolution and ability to change viewing angles, ASTER will produce stereoscopic images, such as 3-D terrain.

QuickBird and Worldview (US)

EarthWatch built two QuickBird 1-meter resolution satellites. EarthWatch later became DigitalGlobe. This constellation is designed to provide a constant digital stream of earth imagery, received and distributed by a global network of ground stations and telecommunications-based distribution networks. As a predecessor to the QuickBird series of satellites, the company had constructed a 3-meter resolution satellite: EarlyBird 1. This satellite was launched successfully on December 24, 1997, but failed on orbit four days later due to a problem with the on-board power system. QuickBird-1 was launched in November 2000 but failed to achieve orbit. QuickBird-2 was launched in October 2001 and is operational. It can provide less than 1m panchromatic and 4m 4-band multispectral images.

WorldView-1 was launched in September 2007. It is the world's only commercial satellite to snap pictures of the Earth at 50-centimeter resolution. Operating at an altitude of 496 kilometers, WorldView-1 has an average revisit time of 1.7 days and is capable of collecting up to 750,000 square kilometers (290,000 square miles) per day of half-meter imagery. The satellite is also equipped with state-of-the-art geolocation accuracy capabilities and exhibits excellent agility with rapid targeting and efficient in-track stereo collection.

WorldView-2 was launched on October 8, 2009. Operating at an altitude of 770 kilometers, WorldView II offers half-meter panchromatic resolution and 1.8-meter multispectral resolution. Added spectral diversity will provide the ability to perform precise change detection and mapping. WorldView II will incorporate the industry standard four multispectral bands (red, blue, green and near-infrared) and will also include four new bands (coastal, yellow, red edge, and near-infrared 2).

WorldView-1 and WorldView-2 are the first commercial satellites to have control moment gyroscopes (CMGs). This high-performance technology provides acceleration up to 10X that of other attitude control actuators and improves both maneuvering and targeting capability. With the CMGs, slew time is reduced from over 60 seconds to only 9 seconds to cover 300km. This means WorldView-2 can rapidly swing precisely from one target to another, allowing extensive imaging of many targets, as well as stereo, in a single orbital pass

IMAGERY CAPABILITIES

Satellite systems are vital for providing earth remote sensing data because of their global coverage and periodic updates. This is especially necessary when gathering information about the environment of a hostile area, remote area or an area we don't have access to.

Multi-spectral imagery can provide information about the environment that can't be detected with the human eye.

Space systems provide broader area coverage than aircraft or UAV systems. One scene from Landsat 5 covers an area of 185km by 170km.

Commercial and civil remote sensing systems provide unclassified imagery. This means it can be used and shared with any coalition partner and does not require stringent handling.

IMAGERY LIMITATIONS

Orbital characteristics have direct influence on the imaging capacity of all spaceborne sensors. Most systems are dependent upon reflected energy from the Sun which makes them effective imagers only during periods of sunlight. In addition, the usefulness of MSI is enhanced for many applications when the images are captured at specific sun elevation angles. Therefore, commercial imaging satellites are typically placed in an orbit described as a "sun-synchronous orbit," allowing the satellites to predictably collect imagery in specified sun angles. Each satellite orbits with specific times established for equatorial crossing; a feature that allows reliable prediction of imaging times at points along its ground track but also prohibits "24 hour" coverage of a specific area. It may be a few days or a few weeks before the satellite covers the area again.

The resolution of the images is not as high as imagery taken from an aircraft or UAV. However, this is rapidly changing with submeter imagery becoming common.

Space environmental effects impact MSI satellites the same as they affect other satellites. Because they are in relatively low earth orbits, they are particularly susceptible to the Earth's atmospheric effects. Terrestrial weather also affects multispectral and IR-only sensor systems. They cannot "see" through clouds, smoke, haze etc.

Multispectral data typically takes significant processing and is converted to prints (photographs, film, and computer compatible storage media) which delays distribution to theater users. Turn around time is normally greater than 24 hours.

Currently, most remote sensing space systems are commercially owned and operated. In times of conflict, these systems could provide products to the enemy. The products from these commercial satellites are for sale unless otherwise restricted. In addition, products can be expensive to purchase.

NATIONAL GEOSPATIAL-INTELLIGENCE AGENCY (NGA)

The National Imagery and Mapping Agency (NIMA) was established 1 October 1996. This agency combined into a single organization, the imagery tasking, exploitation, production and dissemination responsibilities as well as the mapping, charting and geodetic functions of eight separate organizations of the defense and intelligence communities. In October 2003, NIMA was re-designated the National Geospatial-Intelligence Agency (NGA).

NGA incorporates the former Defense Mapping Agency (DMA), Central Imagery Office (CIO) and the Defense Dissemination Program Office in their entirety as well as the mission and functions of the

CIA's National Photographic Interpretation Center (NPIC). Also included in NGA are the imagery exploitation, dissemination and processing elements of the DIA, National Reconnaissance Office (NRO) and the Defense Airborne Reconnaissance Office (DARO).

Besides this major restructuring for support, a new system was put into place for the handling of Production Requirements (PRs) within DoD. This system is called COLISEUM, or the Community On-Line Intelligence System for End-Users and Managers. Eventually, anyone with a PC and having access to the Global Command and Control System (GCCS) will be able to enter PRs via this means.

Although NGA and COLISEUM represent a major change in terms of support to DoD, the process in the Army and the other services for acquiring MSI is still through intelligence channels. As a result, the Collection Manager remains the best starting point for operational and intelligence driven imagery intelligence (IMINT) requirements at the unit level.

PLANNING CONSIDERATIONS

Q. Is there a requirement for IMINT products? Is there an imagery analysis and dissemination infrastructure in place to support rapidly developing operations? New MSI products, because of satellite pass opportunities, may take days to months to obtain. Without a central analysis and dissemination infrastructure, components may have requested and obtained MSI products that are critically required by others.

Q. Are there areas in the supported command's AOR/JOA that require an imagery dissemination infrastructure to be in place for MSI analysis to support rapidly developing operations? MSI products may have to be obtained via contract. Some products may require acquisition lead times to be available when required.

Q. Does the supported command require MSI products for surveillance and strike planning purposes? MSI products typically require lead times (hours to weeks) for delivery which can limit their tactical utility for surveillance and strike. Supporting mechanisms need to be in place to expedite the delivery of MSI products for mission planning, target analysis, terrain categorization, etc.

Q. What is the threat force capability to receive commercial imagery products? Are commercial production facilities or downlink sites in the AO? Commercial imagery products will be available to threat forces in the same manner they are available to friendly forces. They may be direct purchase or through a third party. There may be a commercial imagery downlink and processing station in the threat force country that may want to deny him access to.

FOREIGN RECONNAISSANCE AND SURVEILLANCE

Much of the information about foreign reconnaissance and surveillance systems is classified. In the past, Russia was the only country with considerable reconnaissance and surveillance space systems. China has now become a player in space-based reconnaissance and surveillance along with a few other countries. However, with the advent of space-based commercial 1-meter and sub-meter remote sensing systems, many foreign countries will now have access to reconnaissance capabilities that were typically limited to two or three countries. See the Remote Sensing chapter for more information.

To date, no other country besides the US has a fully operational space-based missile warning system.

Russia

There are several opinions concerning the capability of Russia to continue developing and launching the number of reconnaissance and surveillance systems that it has in the past. The name normally associated with Russian reconnaissance and surveillance satellites is Cosmos (or Kosmos). Recently the number of satellites maintained in orbit has fallen. The gap between the de-orbiting of Cosmos 2320 in 1996 and the launching of Cosmos 2343 in 1997 was unprecedented since the photoreconnaissance satellite program began in 1962.

Reconnaissance Satellites

The Russians are currently flying four types of photoreconnaissance satellites on a semi-regular basis. They have program names such as Yantar, Orlet and Arkon. Most are film return systems although some are now digital data return systems. Typical lifetimes do not exceed one year before they are de-orbited. In the past, the Soviet Union would launch up to 30 photoreconnaissance satellites a year. Over the last few years, there have been periods when Russia has had no record of space reconnaissance capability. The last reported launch was Cosmos 2359 in June 1998. Russia has also started to describe some of their photoreconnaissance satellites in open literature and has sold some of the imagery under the SPIN-2 imaging program.

Electronic Intelligence (ELINT) Satellites

Unlike the photoreconnaissance satellites, many of the ELINT systems are still classified. Russia's capability to maintain a constellation of ELINT satellites has also taken a fall in the past few years. Best know are the ELINT Ocean Reconnaissance Satellites (EORSAT). This program has been in operation since 1974. They have been typically flown to detect, identify and track naval forces that might threaten Russia. Typically two satellites operate simultaneously for a year.

Early Warning Satellites

Russia's early warning space systems for detecting missile attacks is deteriorating. Russia has negotiated with the US during the past year to gain access to some of the US missile warning data. The Russian missile warning space system required several passes over missile site areas and therefore needed several satellites. The current constellation has a gap for several hours. The "Oko" constellation needs at least nine operational satellites in Molniya orbits to provide global overlapping coverage. Russia has not been able to maintain this constellation in the recent years although launches still continue. Russia probably has partial coverage of the US and China. There is also a constellation named Prognoz flown in geosynchronous orbit.

China

Once again, much of the Chinese reconnaissance and surveillance capabilities are classified. China seems to have no long term constellation of military satellites deployed. However, it is possible for the Chinese military to use "civilian" satellites such as the Feng Yun-1 meteorological satellites and Fanhui Shi Weizing (FSW) series of remote sensing satellites. A more capable FSW-3 series will give the Chinese greater capability in the photoreconnaissance arena.

SPACE RECONNAISSANCE AND SURVEILLANCE CAPABILITIES

The prime advantage of reconnaissance and surveillance space systems is their ability to provide worldwide, quick-reaction coverage of areas of interest. They can observe areas that we do not have access to with other sensors. They also enhance planning capabilities, providing updated information regarding enemy force locations. Often the product of a space or terrestrial system can cue a space system to survey an area of interest, enhancing accuracy and reaction times to the user.

Space-based reconnaissance and surveillance systems can provide near-real time information in many cases. TBM warning timelines have decreased significantly with the use of space-based systems.

SPACE RECONNAISSANCE AND SURVEILLANCE LIMITATIONS

The key limitation of these systems is simply based on limitations dictated by the satellite orbit. As discussed with remote sensing and environmental satellites, the LEO orbit gives the best resolution of Earth but it is not a geostationary satellite. The cost of reconnaissance and surveillance satellites currently prohibits a proliferation of satellites as we have with communications satellites. Therefore, the satellites do not cover a particular area 24 hours a day. Commanders and their staffs must know satellite coverage times to effectively enhance their operations.

In addition to the access limitations dictated by the satellite orbit, some satellite systems may be affected by a variety of atmospheric disturbances such as fog, smoke, electrical storms and clouds. These affect the ability of some imaging systems to detect enemy activity and missile launches.

Another problem associated with space-based reconnaissance and surveillance systems is the fact that they are national systems designed to meet the requirements of the POTUS and SECDEF. As such, the priority of effort is at the strategic level and not necessarily at the tactical level of war. Requests for support are based on priority of need.

TBM warning may produce some false events depending on how accurate the commander wants the information. TBM warning systems still experience some correlation problems with other radar systems or with multiple events.

PLANNING CONSIDERATIONS

Reconnaissance and Surveillance

Q. What area do I need information from and what kind of information do I need? The space systems have limitations as to when, where, and what kind of information they can sense.

Q. What kind of information do I need and at what resolution? Can another system such as UAV provide the information I need? The higher the resolution, the longer the tasking will take to get a space system applied against it. The commander must consider whether he has the assets available to get the information he needs before tasking a satellite system.

Q. What priority am I willing to place on the request? How quickly do I need the information? Satellite requests are based on priority of need. Strategic and Operational users frequently have higher priorities than tactical users. It may be hours before your request is processed and then the satellite may not be in position to fulfill your request immediately.

Q. When I get the information, how old is it? Can I use archived information or information that someone else has requested? The commander must always consider the time the information was gathered. If already requested imagery can be used, the time to get the information will be significantly shortened.

TACTICAL EXPLOITATION OF NATIONAL CAPABILITIES (TENCAP)

History

From initial development through to the early 1990s, overhead reconnaissance systems were highly classified and protected by numerous special compartments. Each National Reconnaissance Program (NRP) satellite had its own layer of caveats. The Cold War was on going, and the fact of overhead reconnaissance and the existence of the NRO were classified. The systems were designed to support the POTUS and SECDEF, and Imagery and SIGINT products derived from these sensors remained compartmented to protect the capabilities of the platforms.

TENCAP evolved from an Army recognition that national overhead systems developed for the POTUS and SECDEF had potential to support the tactical warfighter. Throughout the early and mid-'70s, the Army explored the tactical applicability of national systems at corps-level and initiated an office under the Deputy Chief of Staff for Operations (DCSOPS) to develop and field the tools needed to leverage those systems. Congress considered the program so successful that in 1977 it directed the other Services to create similar programs. By the early '80s, TENCAP offices had been established within the Services, the DIA, the National Security Agency (NSA), and the Joint Staff J-3. As the single point conduit into the NRO, the Defense Support Project Office (DSPO) was established under the auspices of OSD, to provide overt oversight of the Defense Reconnaissance Support Program (DRSP) on behalf of the Director, NRO. Through the DRSP, DSPO provided service insight into NRO technologies, facilitated service exploitation of NRO capabilities, and funneled service funding into the NRO for DoD unique capabilities. Throughout the early years of TENCAP, national overhead reconnaissance systems remained focused on support to the national community. Although the DoD TENCAP community made tremendous strides in leveraging national systems for the tactical user, it wasn't until after Operations DESERT SHIELD/DESERT STORM that the military customer really became a critical consideration when evaluating requirements for future NRO systems. The DRSP was renamed the Defense Space Reconnaissance Program (DSRP) in 1994 as an element of the Joint Military Intelligence Program (JMIP).

With decompartmentalization and declassification within the overhead reconnaissance community over the past 5 years, roles and functions of TENCAP-related organizations have evolved. Several services have reorganized to encompass both "white" and "black" space responsibilities. Support to military operations has become a major driver in the requirements process for new overhead architectures. The recent openness of the NRO has led to better dialogue with the military customers across the board. Requirements for emerging NRP satellites are now addressed by the Joint Requirements Oversight Council (JROC), a situation totally unheard of just three years ago. Even though much has changed in the way the DoD and the national community do business, the value placed by the Services on their TENCAP Programs is reflected by the fact the funding for these efforts has remained relatively consistent in a period of declining resources.

Army TENCAP

The Army Space Program Office (ASPO) has been developing, acquiring, fielding, and maintaining Army TENCAP systems for 25 years. It is the largest service TENCAP Program, and is unique in its life cycle approach to TENCAP. From the program's inception, the Army recognized the need to normalize TENCAP like any other weapons system. By incorporating it as an element within the US Army Training and Doctrine Command's (TRADOC) Intelligence and Electronic Warfare (IEW) Battlefield Operating System (BOS), TENCAP systems have been fully integrated into the division, corps, and echelon above corps (EAC) intelligence infrastructure. With the consolidation of space functions in 1994, ASPO transferred to the US Army Space and Missile Defense Command (SMDC), and the SMDC Acquisition Center. In 2003, the mission of the Program Executive Office (PEO) for Air and Missile Defense was expanded to include the Army Space Program Office. The Army TENCAP program baseline

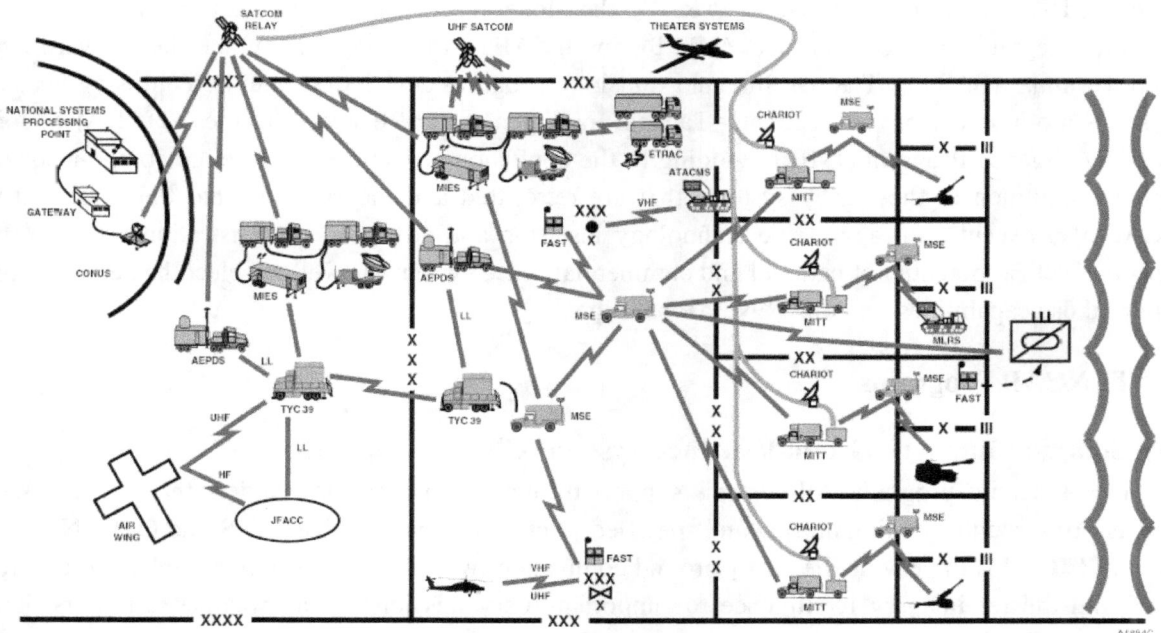

Figure 10-6: Army TENCAP Architecture

has undergone a major transition, with migration to a modular, downsized baseline system called the

Tactical Exploitation System (TES). The TES is a split-based, scaleable, single-integrated, multi-

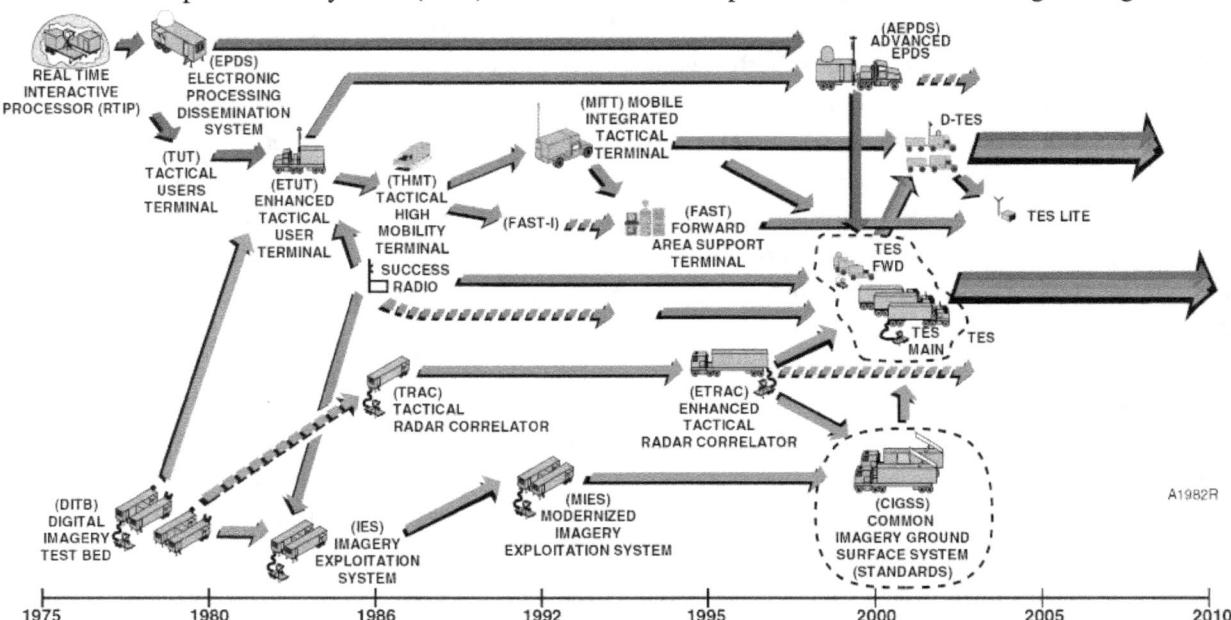

Figure 10-7: Army TENCAP Evolution

disciplined, all-source, preprocessing family of systems.

Legacy systems include the Advanced Electronic Processing and Dissemination System (AEPDS), the Enhanced Tactical Radar Correlator (ETRAC), the Modernized Imagery Exploitation System (MIES), the Mobile Integrated Tactical Terminal (MITT), and the MITTs functional twin, and the Forward Area Support Terminal (FAST). The baseline and transition program descriptions can be confusing because the systems are in either of two accounts--Tactical Intelligence and Related Activities (TIARA) or Joint Military Intelligence Program (JMIP). Adding to the confusion are numerous baseline components and subsystems common to the legacy systems that are retrofitted and incorporated into the TES. Army TENCAP also executes an aggressive technology insertion and advanced demonstration program that exploits the tactical potential of national and commercial space systems as well as select theater assets and integrates those capabilities into the TENCAP baseline.

Army TENCAP Programs

Army Common Imagery Ground/Surface System (CIGSS - components)

CIGSS is a joint program. All systems support theater combatant commanders through the Army component or through standing unified and specified joint command's CONPLANS and OPLANS. The purpose of CIGSS is to provide the Army ground commander with timely, assured, reliable, and accurate national and theater imagery intelligence to support military missions, in all-weather, 24-hours daily. Programmatically, TES is the Army component of CIGSS joint service imagery baseline.

Migration to the Army common imagery ground station includes continued engineering and technology insertion upgrades to the processors, and processing tools. Army will maintain the Common

Imagery Processor (CIP) baseline program capability to receive and process data from all DOD imagery collection platforms. Efforts include: ASARS Improvement Program (AIP) sensor implementation; automatic target recognition (ATR) projects, like the Semi-automated IMINT Processor (SAIP) ACTD; upgrades to the common data link (CDL) for the TES.

Tactical Exploitation System (TES)

TES is the Army's objective TENCAP system for the 21st century, and it replaces the AEPDS, the ETRAC, and the MIES. TES combines all TENCAP functionality into a single, integrated, scaleable system, designed for split-based, disbursed operations. TES development includes joint CIGSS components, CIP development, compliant and compatible communications systems and sub-systems, and acquisition of a TES system (Littoral Surveillance System (LSS)) by the Navy Reserves.

TES is a multi-disciplined processing system that will provide the tactical and operational commander with timely and assured access to imagery and signals intelligence from national, theater, and select organic tactical collectors. TES will perform the preprocessor functions for forward and main elements, as well as the division-TES (D-TES) and TES-light configurations. TES is designed for split-based and extended battlespace operations. The TES baseline consists of the All-Source Analysis System (ASAS), the common ground station (CGS), and the Digital Topographic Support System (DTSS).

The TES forward is a highly mobile, HMMWV-based system that can be multiple mission configured in one of five options to meet the tactical situation. This is a fly-away (C-130 drive-on drive-off, quick set-up tear-down, rapid deployment, early-entry) system.

The TES main is housed in vans and will operate from sanctuary, or can be deployed if the United States enters into sustained theater military operations, like Operations DESERT SHIELD/STORM.

The D-TES and TES-light will be a downsized TES forward configuration, fielded to the current Mobile Integrated Tactical Terminal (MITT) and Forward Area Support Terminal (FAST) force baseline.

The Army acquisition objective is for six TES (XVIII ABN Corps, 513th MI BDE, III Corps, 501st BDE, V Corps, and I Corps). The Army procurement objective will include a mix of full and TES (-) configurations to be acquired and fielded over the POM. D-TES and TES-light will be fielded at the division and separate BDE level. The basic order of issue (BOI) requirement for TES is one Main or Forward per MI BDE at Corps and EAC, and one DTES or TES-light per MI BN at division and MI CO at separate maneuver BDE.

TES-LITE

Fielded in the 2006 TES-LITE accepts, correlates, and integrates SIGINT reports from national, theater, and corps collection and dissemination assets to include the AEPDS. Receives, integrates and disseminates national and theater imagery from various processors and sources. Tactical Exploitation System-Light (TES-LITE) incorporates the functionality of the FAST. The TES-LITE is a transportable, modular, survivable, stand-alone, UNIX system. It provides MITT functionality in transit cases.

SIGINT & IMINT Preprocessing Architecture (S&IPA)

The S&IPA provides a framework for the TES and the Guardrail Common Sensor (GRCS) Integrated Processing Facility (IPF) and its follow-on, the Aerial Common Sensor (ACS) Ground Processing Facility (GPF) to become interoperable, achieve selective commonality of processors and methods, and to operate as an integrated system-of-systems. The S&IPA will also provide the flexibility for split-based operations, that support deployable and deployed forward and main headquarters elements of the joint task force or ground component commander (at corps or detached to the division or brigade echelons coincident with the contingency operation). The S&IPA reflects a transition from autonomous TENCAP and GRCS IPF system operations to a state of interoperability and selective commonality for TES and ACS GPF. The S&IPA will support national security objectives and assist in the planning and execution of strategic, operational, and tactical deployments. It becomes part of contingency operations as an element of an Army, joint, or combined force for stability and support operations or for war across the spectrum of conflict.

The S&IPA will provide a single preprocessing resource and focus for the commander, G-2, and G-3. It will leverage the synergy of collocated analysts and their methods, techniques, and tip-offs. These characteristics will achieve a timely coherent data stream to the G-2 and G-3, so that the data may be fused to form a common picture of the battlefield for dissemination. The S&IPA supports dynamic tasking, retasking, and synchronized collection, processing, and exploitation of national, theater, and organic platforms and sensors.

Eagle Vision II (EVII)

EVII is a proof-of-concept experiment to help determine the utility of timely civil and commercial imagery as a responsive combat enabler. The goals are: 1) to assess the usefulness of commercial high-resolution imagery to support military operations; 2) to develop technical and operational architectures for use with existing military infrastructure, to determine best fit (separate system or integrated into existing or developing systems); 3) formulate concepts for dissemination of unclassified imagery within joint and collation force structures and architectures.

EVII is a joint Army and NRO effort. Operational demonstration and evaluation will be conducted by the Army. An EVII integrated product team (IPT) is the mechanism by which changes to the system and the concept of experiment (CONEX) are proposed, evaluated, and prioritized by the CCB. The ASPO Project Officer chairs the EVII IPT, which includes representatives from the NRO, SMDC, HQDA, USARSTRAT, TRADOC, and TRADOC Battle Labs. The NRO will chair the CCB, which include ASPO, TEC, and ARSTRAT. Eagle Vision II has been incorporated as an Initiative in JCS Special Project SOUTHERN EYE.

Due to funding and schedule constraints, a phased development plan is being executed to provide this experimental capability. The EVII program is structured into two phases. Phase I will develop, build, test, and operate the prototype mobile satellite commercial imagery ground receive and processing system. Phase Ia will deliver a system capable of ingesting data from SPOT 1, 2, and 4 and RADARSAT 1. Phase 1b will provide a field upgrade to add the capability to ingest data from one US-built high-resolution system. Additionally, Phase 1 will include a Joint Collection Management Tool (JCMT) based EV Mission Manager (EVMM) that will provide a common commercial imagery planning tool. Phase II is the planned incorporation of one additional high-resolution system and an upgrade and license to receive Landsat 7.

The Army has a continuing need to evaluate evolving commercial imagery operations against specific Army criteria. Inherent to this effort is the need to study, test, and evaluate the capability and performance of individual commercial imaging satellite systems. EVII will test the entire tasking, processing, exploitation, and dissemination (TPED) for commercial imagery. The results from EVII will help define the role of commercial imagery for strategic, operational, and tactical users. EVII will provide a deployable mobile ground station that is not dependent on a foreign controlled downlink and processing facility.

EVII will be certified for air transport on C-130 or larger military airlift. The system will consist of a diesel tractor, a 34-foot double-expandable semi-trailer, and a separate, towable, 5.4 meter antenna. The dual push-out design of the van provides extra workspace, and the system is equipped with a remote operating position (ROP) that can be set up in the supported organizations operations center. The system uses a COTS-GOTS open architecture.

The EVII network and communications subsystem will provide connectivity between the multiple internal computer systems, and to external military and commercial networks. The network includes an asynchronous transfer mode (ATM) for large data transfers and an Ethernet. The subsystem supports both unclassified commercial and classified voice and fax data communications. It will employ standard area communications, and commercial communications capable of worldwide commercial connectivity, used as primary military backup, and for interfacing with commercial vendors.

Appendix A
Abbreviations and Acronyms

3D	Three Dimensional
3Y	Additional Skill Indicator for Space Activities
AAN	Army After Next
AAMDC	Army Air and Missile Defense Command
ABCS	Army Battle Command System
ABL	Airborne Laser
ABM	Antiballistic Missile
ACOM	Atlantic Command
AF	Air Force
AFB	Air Force Base
AFFOR	Air Force Forces
AFS	Air Force Station
AFSATCOM	Air Force Satellite Communications System
AFSCN	Air Force Satellite Control Network
AFSPACE	Air Force Component to USSTRATCOM (14th Air Force)
AFSPC	Air Force Space Command
AJ/AS	Anti-Jam/Anti-Spoof
ALCOR	ARPA Lincoln C-Band Observation Radar
ALRPG	Army Long Range Planning Guidance
ALERT	Attack, Locate and Early Report to Theater
ALTAIR.	ARPA Long-range Tracking and Identification Radar
AOI	Area of Interest
AOR	Area of Responsibility

ARFOR	Army Forces
ARPA	Advanced Research Projects Agency (Now DARPA)
ARSST	Army Space Support Teams
ARSTRAT	Army Forces Strategic Command
ARTS	Automated Remote Tracking System
A-S	Anti-Spoofing
ASARS	Advanced Synthetic Aperture Radar System
ASAS	All Source Analysis System
ASAT	Anti-Satellite
ASCC	Alternate Space Control Center
ASEDP	Army Space Exploitation Demonstration Program
ASEWG	Army Space Executive Working Group
ASI	Additional Skill Indicator
ASMP	Army Space Master Plan
ASP	Army Space Policy
ASPO	Army Space Program Office
ASSC	Army Space Support Cell
ATD	Advanced Technology Demonstration
ATM	Asynchronous Transfer Mode
ATMDE	Army Theater Missile Defense Element
AV 2010	Army Vision 2010
AWE	Advanced Warfighting Experiment
AWS	Advanced Wideband System
BCBL	Battle Command Battle Lab
BDA	Battle Damage Assessment
BFT	Blue Force Tracking

BLOS	Beyond Line of Sight
BM	Battle Manager
BMC	Ballistic Missile Center/Battle Management Cell
BMC3	Ballistic Missile Command, Control and Communications
BMD	Ballistic Missile Defense
BMDO	Ballistic Missile Defense Office
C2	Command and Control
C4I	Command, Control, Communications, Computers, and Intelligence
C4ISR	Command, Control, Communications, Computers, Intelligence, Surveillance and Reconnaissance
CALL	Center for Army Lessons Learned
CASCOM	Combined Arms Support Command
CCD	Coherent Change Detection
CC&D	Camouflage, Concealment and Deception
CCIS	Civil/Commercial Imagery Systems
CECOM	Communications and Electronic Command
CENTCOM	Central Command
CEP	Circular error of probability
CFJO	Concept for Future Joint Operations
CG	Commanding General
CGS	Common Ground Station
CIA	Central Intelligence Agency
CIGSS	Common Imagery Ground/Surface System
CIP	Common Imagery Processor
CJCS	Chairman, Joint Chiefs of Staff
CJTF	Combined Joint Task Force

Cm	centimeter
CMAFB	Cheyenne Mountain Air Force Base
CMAS	Cheyenne Mountain Air Station
CMO	Commercial Satellite Communication Initiative Management Office
CMOC	Cheyenne Mountain Operations Center
CNO	Chief of Naval Operations
COA	Course of Action
COCOM	Combatant command
COIL	Chemical Oxygen-Iodine Lasers
COMAFSPACE	Commander, Air Force Space Command (14th AF)
COMARSTRAT	Commander Army Forces Strategic Command
COMM	Communications
COMSAT	Communications satellite
COMSEC	Communications Security
CONOPS	Concept of Operations
CONR	CONUS NORAD Region
CONUS	Continental United States
COP	Common Operating Picture
CoS	Control of Space/ Chief of Staff
COTS	Commercial-Off-The-Shelf
CSLA	Commercial Space Launch Act
CSCI	Commercial Satellite Communications Initiative
DAMA	Demand Assigned Multiple Access
DARPA.	Defense Advanced Research Projects Agency
DAWE	Division Advanced Warfighter Experiment
DNI	Director of National Intelligence

D & D	Denial & Deception
DDS	Defense Dissemination System
DDNI	Deputy Director of National Intelligence
DDL	Direct Downlink
DE	Directed Energy
DEW Line	Distant Early Warning Line
DGPS	Differential GPS
DISA	Defense Information Systems Agency
DISN	Defense Information Systems Network
DITP	Discriminatory Interceptor Technology Program
DMA	Defense Mapping Agency
DMS	Defense Message System
DMSP	Defense Meteorological Satellite Program
DNRO	Director of the National Reconnaissance Office
DOC	Department of Commerce
DOCC	Deep Operations Coordination Cell
DoD	Department of Defense
DOT	Department of Transportation
DSCS III	Defense Satellite Communications System Phase III
DSCSOC	DSCS III Operations Center
DSNET	Defense Security Network
DSP	Defense Support Program
DTD	Digital Terrain Data
DTES	Division Tactical Exploitation System
DTED	Digital Terrain Elevation Data
DTSS	Digital Topographical Support System

DTSS-D	Digital Topographical Support System-Division
DTSS-H	Digital Topographical Support System-Heavy
DTSS-L	Digital Topographical Support System-Light
DUSD(Space)	Deputy Under Secretary of Defense for Space
DWSW	Deployable Weather Satellite Workstation
EAC	Echelons Above Corps
ECB	Echelons Corps and Below
EELV	Evolved Expendable Launch Vehicle
EAM	Emergency Action Message
EGI	Embedded GPS Inertial
EHDR	Extreme High Data Rate
ELINT	Electronic or Electromagnetic Intelligence
ELV	Expendable Launch Vehicle
EMP	Electro Magnetic Pulse
EMUT	Enhanced Manpack UHF Terminal
E/O	Electro-Optical
EOSAT	Earth Observation Satellite
EPLGR	Enhanced PLGR
EPLRS	Enhanced Position, Location, Reporting System
ERM	Earth Resource Monitoring
ERS	Earth Resourcing Satellite
ESA	European Space Agency
ETRAC	Enhanced Tactical Radar Correlator
ETUT	Enhanced Tactical Users' Terminal
EUCOM	European Command
EW	Electronic Warfare

EWE	Early Warning Experiment
EWR	Early Warning Radar
FA 40	Functional Area 40 (Space Operations Officer)
FAA	Federal Aviation Administration
FAR	Federal Acquisition Regulation
FAST	Forward Area Support Terminal
FCC	Federal Communications Commission
FDIC	Force Development Integration Center
FDO	Flexible Deterrent Options
FEP	FLTSAT EHF Program
FLTSAT	Fleet Satellite
FLTSATCOM	Fleet Satellite Communication System
FOC	Future Operational Capability
FOV	Field of View
FY	Fiscal Year
FYDP	Five Years Defense Program
G-B	Grenadier-Beyond Line of Sight Ranging and Tracking System
GBE	Ground-Based EMP
GBI	Ground-Based Interceptor
GBL	Ground-Based Laser
GBS	Global Broadcast Service
GCCS	Global Command and Control System
GCN	Ground Communications Network
GCSS	Global Command Support System
GDIN	Global Defense Information Network
GE	Global Engagement

GEO	Geosynchronous Earth Orbit/Geostationary Earth Orbit
GEODSS	Ground-Based Electro-Optical Deep Space Surveillance
GIS	Geographic Information System
GLOBALSTAR	Commercial Name of System
GMF	Ground Mobile Forces
GMPCS	Global Mobile Personal Communications Satellite
GMS	Geostationary Meteorological Satellite
GOES	Geostationary Operational Environmental Satellite
GOMS	Geostationary Orbiting Meteorological Satellite
GPS	Global Positioning System
Grenadier-BRAT	Grenadier-Beyond Line of Sight Range and Tracking System
HA-UAV	High Altitude Unmanned Aerial Vehicle
HDR	High Data Rate
HEMP	High Altitude Electro-Magnetic Pulse
HEO	Highly Elliptical Orbit
HQDA	Headquarters, Department of the Army
HSI	Hyper-Spectral Imagery
IADS	Integrated Air Defense System
IBS	Integrated Broadcast Service
IC	Intelligence Community
ICBM	Intercontinental Ballistic Missile
ID	Identification/Identify
IFSAR	Interferometric Synthetic Aperture Radar
IGEB	Interagency GPS Executive Board
IMETS	Integrated Meteorological Systems
IMINT	Imagery Intelligence

INFOSEC	Information Security
INSCOM	Army Intelligence Security Command
IO	Information Operations
I/O	Input/output
IPB	Intelligence Preparation of the Battlespace/Battlefield
IPL	Integrated Priority List
IR	Infrared
IRS	India Remote Sensing satellite
ISR	Intelligence, Surveillance, and Reconnaissance
ISS	International Space Station
ISSN	Integrated Space Surveillance Network
ITSO	International Telecommunication Satellite Organization
ITW/AA	Integrated Tactical Warning and Attack Assessment
JAOC	Joint Air Operations Center
JCAPA	Joint/Combined Arms Precision Attack
JCS	Joint Chiefs of Staff
JFACC	Joint Force Air Component Commander
JFLCC	Joint Force Land Component Commander
JFMCC	Joint Force Maritime Component Commander
JFS	Joint Feasibility Study
JFSCC	Joint Force Space Component Commander
JP	Joint Publication
JPO	Joint Program Office
JROC	Joint Requirements Oversight Council
JRSC	Jam Resistant Secure Communications
JSIC	Joint Space Intelligence Center

JSIMS	Joint Simulation System
JSMB	Joint Space Management Board
JSpOC	Joint Space Operations Center
JSST	Joint Space Support Team
JSTARS	Joint Surveillance Target Attack Radar System
JTAGS	Joint Tactical Ground Station
JTF	Joint Task Force
JTF/CC	Joint Task Force Component Commander
JTIDS	Joint Tactical Information Distribution System
JV 2010	.Joint Vision 2010
Kbps	Kilobits Per Second
KE	Kinetic Energy KE
KE ASAT	Kinetic Energy Anti-satellite
KHz	KiloHertz
km	Kilometer
KW	Kinetic Weapon
LADAR.	Laser Detection and Ranging
LAN	Local Area Network
LANDSAT	Land Satellite
LCC	Land Component Commander
LDR	Low Data Rate
LEO	Low Earth Orbit
LEOCOMM	Low Earth Orbit Communications
LIDAR	Laser Identification Detection and Ranging
LNO	Liaison Officer
LO	Low Observable

LoD	Launch on Demand
LOS	Line of sight
LPD	Low Probability of Detection
LPI	Low Probability of Intercept/Launch and Predicted Impact
LWIR	Longwave IR
m	meter
MABS	Missile Alert Broadcast System
MAGR	Miniaturized Airborne GPS Receiver
MAP	Mission Area Plan
MAPS	Mission Area Planning System
MASINT	Measurement and Signals Intelligence
MAST	Meteorological Automated Sensor and Transceiver
MCG	Mapping, Charting, and Geodesy
MCS	Maneuver Control System
MDR	Medium Data Rate
MDSTC	Missile Defense and Space Technology Center
MEO	Mid Earth Orbit
METL	Mission Essential Task List
METEOSAT	Meteorological Satellite
METOC	Meteorological and Oceanographic
METT-T	Mission, Enemy, Own Troops, Terrain, and Tim
MGES	Mobile Gateway Earth Station
MGS	Mobile Ground Station
MIC	MILSATCOM Iridium Communications
MIDAS	Miniaturized Data Acquisition System

MIES	Modernized Imagery Exploitation System
MILSATCOM	Military Satellite Communications
MIL-STD	Military Standard
MIRACL	Mid IR Advanced Chemical Laser
MIST	Multiband Integrated Satellite Terminal
MITT	Mobile Integrated Tactical Terminal
MMBL	Mounted Maneuver Battle Lab
MOA	Memorandum of Agreement
MSE	Mobile Subscriber Equipment
MSI	Multispectral Imagery
MSS	Mobile Satellite Services/Multispectral Scanner
MSTS	Multi-Source Tactical System
MWC	Missile Warning Center
NASA	National Aeronautics and Space Administration
NATO	North Atlantic Treaty Organization
NAV	Navigation
NAVFOR	Naval Forces
NAVSOC	Naval Satellite Operations Center
NAVSPACECOM	Naval Space Command
NAVSPASUR	Naval Space Surveillance System
NAVWAR	Navigation Warfare
NBC	Nuclear, Biological, and Chemical
NCO	Noncommissioned Officer
NEMSS	National Environmental Monitoring Satellite System
NETWARCOM	Network Warfare Command (Naval)
NGA	National Geospatial-Intelligence Agency

NGO	Non-Governmental Organization
NIIRS	National Imagery Interpretability Rating Scale
NIMA	National Imagery and Mapping Agency
NIR	Near Infrared
NMD	National Missile Defense
NMS	National Military Strategy
NNSOC	Naval Network and Space Operations Command
NOAA	National Oceanic and Atmospheric Administration
NORAD	North American Aerospace Defense Command
NPOESS	National Polar Orbiting Environmental Satellite System
NRO	National Reconnaissance Office
NRT	Near Real-Time
NSA	National Security Agency
NSC	National Space Council/National Security Council
NSP	National Space Policy
NSSMP	National Security Strategy Master Plan
NSS	National Security Strategy
NSSA	National Security Space Architect
NTC	National Training Center
NUDET	Nuclear Detonation
NVG	Night Vision Goggles
OCONUS	Outside of Continental United States
ODA	Optical Data Analysis
ODOC	Objective DSCS III Operations Center
O&M	Operations and Maintenance
OLS	Operational Linescan System

OOTW	Operations Other Than War
OPSEC	Operations Security
ORD	Operational Requirement Document
OSA	Office of the Space Architect
OSD	Office of the Secretary of Defense
OSF	Operational Support Facility
P-code	Precision Code
PACOM	Pacific Command
PAFB	Peterson Air Force Base
PAM	Pamphlet
PCP	Precision Code Protected
PCS	Personal Communications Services
PD	Presidential Directive
PDD	Presidential Decision Directive
PLGR	Precision Lightweight GPS Receiver
POC	Point of Contact
POES	Polar Orbiting Environmental Satellite
POS/NAV	Position/Navigation
POTUS	President of the United States
PPBE	Planning, Programming, Budgeting and Execution System
PPS	Precision Positioning Services
PPS-M	Precision Positioning Services-Security Module
PSM	Portable Space Model
PVO	Private Volunteer Organization
R&D	Research and Development
RBV	Rapid Battlefield Visualization

RDEC	Research, Development, and Engineering Center
RDT&E.	Research Development Test and Evaluation
REG	Regulation
RFI	Radio Frequency Interference
RIDSN	Radar Imaging and Deep Space Network
RISTA	Reconnaissance, Intelligence, Surveillance, and Target Acquisition
RLV	Reusable Launch Vehicle
R&S	Reconnaissance and Surveillance
RT	Real Time
SA	Situational Awareness/Selective Availability
SAGR	Standalone Air GPS Receiver
SAR	Synthetic Aperture Radar
SATCOM	Satellite Communications
SATRAN	Satellite Reconnaissance Advance Notice
SBC	Synthetic Battle Center
SBE	Synthetic Battlefield Environment
SBI	Space-Based Interceptor
SBIRS	Space-Based Infrared System
SBIRS-H	SBIRS- High
SBJ	Space-Based Jammer
SBL	Space-Based Laser
SBP	Space-Based Platform
SBR	Space-Based Radar
SCAMP	Single Channel Anti-Jam Man Portable
SCC	Space Control Center
SCI	Sensitive Compartmented Information

SECARMY	Secretary of the Army
SECDEF	Secretary of Defense
SEL	Space Education and Literacy
SEP	Spherical Error Probability
SHF	Super High Frequency
SIAM	Space and Information Analysis Model
SID	Secondary Imagery Dissemination
SIGINT	Signals Intelligence
SINCGARS	Single Channel Ground and Air Radio System
SIPRNET	Security Internet Protocol Routing Network
SLBM	Sea-Launched Ballistic Missile
SLEP	Service Life Extension Program
SLGR	Small Lightweight GPS Receiver
SLV	Space Launch Vehicle
SMART-T	Secure Mobile Anti-Jam Reliable Tactical Terminal
SMDBL	Space and Missile Defense Battle Lab
SMDC	Space and Missile Defense Command
SOB	Space Order of Battle
SOC	Space Operations Center
SOCOM	Southern Command
SOF	Special Operations Forces
SOI	Space Object Identification
SOLGR	Special Operations Lightweight GPS Receiver
SOS	System of Systems
SPACECOM	Space Command
SPAWAR	Space and Naval Warfare Systems Command

SPIN	SATCOM Planning Information Network
SPIRIT	Special Purpose Integrated Remote Intelligence Terminal
SPOC	Space Operations Center (USSTRATCOM)
SPOT	Satellite Probatoire d'Observation de la Terre
SPS	Standard Positioning Service
SRBM	Short Range Ballistic Missile
SSDC	Space and Strategic Defense Command
SSN	Space Surveillance Network
STS	Space Transportation System
STAR-T	SHF Tri-band Advanced Range Extension Terminal
STEP	Standardized Tactical Entry Point
STRATCOM	Strategic Command
STO	Science and Technology Objective
STT	Small Tactical Terminal
SUCCESS	Synthesized UHF Computer Controlled Equipment Subsystem
SWC	Space Warfare Center
SWIR	Shortwave Infrared
TAC v2	Tactical Communication Protocol version 2
TACDAR	Tactical Detection and Reporting
TACON	Tactical Control
TBM	Theater Ballistic Missile
TD	Technology Demonstration/Training Development
TDDS	TRAP (TRE- Tactical Receive Equipment—and Related Applications) Data Dissemination System
T&E	Test and Evaluation
TEC	Topographical Engineering Center

TENCAP	Tactical Exploitation of National Capabilities
TEL	Transporter Erector Launcher
TES	Theater Event System/Tactical Exploitation System
THAAD	Theater High Altitude Air Defense
THEL	Tactical High Energy Laser
TIBS	Tactical Information Broadcast Service
TM	Thematic Mapper
TMD	Theater Missile Defense
TOC	Tactical Operations Center
TP	TRADOC Pamphlet
TRANSCOM	Transportation Command
TRAP	Tactical and Related Applications
TRADOC	Training and Doctrine Command
TRE	Tactical Receive Equipment
TRI-TAC	Tri-Service Tactical Communications
TT&C.	Telemetry, Tracking and Commanding
TTP	Tactics, Techniques, and Procedures
TW/AR.	Threat Warning/Attack Reporting
UAV	Unmanned Aerial Vehicle
UCP	Unified Command Plan
UFO	UHF Follow-on
UHF	Ultra High Frequency
US	United States
USA	United States Army
USAF	United States Air Force
UAV	Ultralight Aerial Vehicle

USD (A & T)	Under Secretary of Defense for Acquisition and Technology
USEUCOM	United States European Command
USMC	United States Marine Corps
USN	United States Navy
USI	Ultra-Spectral Imagery
USSPACECOM	United States Space Command (became part of USSTRATCOM in 2002)
USSTRATCOM	United States Strategic Command
WAAS	Wide Area Augmentation System
WAGE	Wide Area GPS Enhancement
WEFAX	Weather Facsimile
WMD	Weapons of Mass Destruction
WRAP	Warfighting Rapid Acquisition Program
WSMC	Western Space and Missile Center
WTEM	Weather, Terrain, and Environmental Monitoring
WIN-T	Warfighter Information Network-Terrestrial
WWMCCS	Worldwide Military Communication Command and Control System
WWW	World Wide Web

Appendix B
Glossary of Terms

Access   A channel allocated to a specific user for a specific period of time.

Anti-Jam   Resistant to signals which would interfere with reception of the desired communications. Jamming is not necessarily hostile and an be caused by unintentional use of the system or improperly tuned equipment.

Anti-satellite (ASAT)   Any weapon designed to destroy satellites.

Apogee   The point in a satellite's orbit where it is farthest from the Earth and its velocity is slowest.

Argument of Perigee   The orbital element of the angular measurement from the right ascension of the ascending node along the orbital path, in the direction of satellite motion, to the perigee point.

Ascending Node   The point where a satellite crosses the equator in a south to north direction.

Assured access   Allocation of the necessary satellite resources to form communication slinks or networks when needed throughout the strategic, operational and tactical areas of operations.

Asynchronous Transfer
Mode (ATM)   This is the new form of super-fast packet switching. In the 21st century ATM networks will operate at speeds of gigabits per second.

Atmospheric drag   Resistive forces caused by gases in the atmosphere acting on an orbiting satellite.

Band (channel)   As it refers to remote sensing, a band is a slice of wavelengths from the electromagnetic spectrum.

Bandwidth   The width of a given band or spectrum of frequencies of interest, expressed in hertz. The lowest usable frequency subtracted from the highest usable frequency for a communication channel gives its bandwidth. Generally, higher bandwidth channels have greater capacity to convey signals modulated with higher data rates of information.

Baud	The number of pulses per second or the number of times per second that a signal on a communication circuit changes.
Beamwidth	The angle of the conical shaped beam that an antenna radiates. Large antennas have narrower beamwidths and can pinpoint satellites in space or dense traffic areas on the Earth more precisely. Tighter beamwidths deliver higher levels of power and thus greater communications performance.
Bent Pipe	A non-regenerative non-processed channel that does nothing to the signal received by the satellite, except to relay it toward Earth.
Bus	Everything on a satellite except the payload itself. Normally includes the structural frame, TT&C subsystems, power, attitude control, thermal management systems, etc.
Channel	As it refers to communications, a frequency bandwidth available as a communications path. A channel will normally be the operating frequency and bandwidth of a satellite transponder.
Circular Orbit	Any orbit that has an eccentricity of zero. Since all satellites are subject to perturbations, no orbit can achieve and maintain an eccentricity of exactly zero. Common practice is to call orbits with very low eccentricities circular.
Constellation	A number of like satellites that are part of a system. Satellites in a constellation generally have the same type orbit, although that is not a requirement.
Control Segment	That portion of a space system that controls the satellite platform, payload, and network. This segment provides for station-keeping, orbital changes, attitude and stabilization changes, and other satellite maintenance and housekeeping activities.
Countermeasure	Any action or measure employed by a threatened or targeted system to avoid detection, destruction, or neutralization.
Corona	The outermost layer of the Sun's atmosphere. The corona is very hot, up to 1-1.5 million degrees centigrade, and is the source of the *solar wind*.
Coronal mass ejection	A huge cloud of hot plasma, expelled sometimes from the Sun. It may *accelerate ions* and *electrons*, and may travel through interplanetary

space as far as the Earth's orbit and beyond it, often preceded by a *shock front*. When the shock reaches Earth, a *magnetic storm* may result.

Coverage	The portion of the earth's surface over which satellite services are provided.
Crosslink	A communications link between satellites, usually a microwave, millimeter wave, or laser signal with a narrow beamwidth. Crosslinked satellites provide connectivity to satellites that are out of view of ground stations.
DAMA	Demand Assigned Multiple Access. An efficient method of managed channel access allowing the sharing of one or more channels by multiple users. A control system providing satellite access to customers on a priority and need basis. DAMA reduces the amount of unused (wasted) satellite channel availability time.
Decay	Uncontrolled reentry of a satellite as a result of atmospheric drag.
Dedicated sensor	Sensor owned and operated by AFSPACECOM whose primary mission is space surveillance.
Descending node	A point at the intersection of the equatorial plane and the orbital plane of a southbound satellite.
Digital image	an image that has been placed in a digital file with brightness values of picture elements (pixels) representing brightness of specific positions within the original scene. The original scene may be the Earth as digitized by sensors in space or it may be a picture scanned by a desktop or other variety of scanner.
Downlink	A communications channel from a satellite to an Earth station.
Eccentricity	The amount by which an orbit varies from perfectly circular. As this value approaches one, the shape becomes more elongated. As this value approaches zero, the shape becomes more circular.
Electromagnetic field	The regions of space near *electric currents*, magnets, broadcasting antennas etc., regions in which electric and magnetic forces may act. Generally the EM field is regarded as a modification of space itself, enabling it to store and transmit energy.

Electromagnetic perturbation	Magnetic drag caused by the interaction of the Earth's magnetic field and the satellite's electromagnetic field.
Electromagnetic radiation	Energy transfer in the form of electromagnetic waves or particles that propagate through space at the speed of light.
Electromagnetic spectrum	The entire range of electromagnetic radiation. The spectrum usually is divided into seven sections. From the longest wavelengths to the shortest: radio, microwave, infrared, visible, ultraviolet, x-ray, and gamma ray radiation
Elliptical orbit	Any orbit that is not perfectly circular is elliptical. The term is normally used to describe orbits that vary significantly from circular.
Ephemeris data	Information needed to establish a link to a satellite, including look angles in elevation and azimuth.
Epoch Time	A particular instant in time for which satellite measurements of position are made.
Equatorial Orbit	Satellites in equatorial orbit are, by definition, inclined at zero degrees. All geostationary orbits are also equatorial orbits.
Footprint	A satellite's footprint is that area on the Earth's surface covered by the satellite antenna's beam pattern or within the field of view of the satellite's transmitters or sensors.
Frequency	How frequently an electromagnetic signal completes one complete cycle in one second, expressed in Hertz.
Frequency Hopping	Discrete jumping of a signal's transmitted frequency over time. Used to counter jamming or other interference.
Full Duplex	Able to communicate both ways (transmit and receive) simultaneously
Gapfiller	A commercial satellite capability leased by the Navy to provide UHF communications. Originally built to fill the gap preceding the FLTSAT launches, they still provide SATCOM service for DOD. Will be replaced by UHF Follow-on. (UFO utilizes Gapfiller frequencies.)
Gateway	A ground station that acts as a relay between satellites in a system or a link between the satellite and entry into the terrestrial communications network.

Geostationary Orbit	A special type of geosynchronous orbit which is nearly circular, has an inclination of approximately zero degrees, and a period of one day. A satellite in geostationary orbit appears to remain fixed in the sky above the equator when observed from the earth's surface. A typical geostationary orbit has, at an altitude of approximately 22,300 miles over the equator, an orbital period of 24 hours thus coinciding with the rotation period of the Earth.
Geosynchronous orbit (GEO)	Any orbit with an orbital period of one day. A satellite in a geosynchronous orbit does not necessarily appear to be stationary in the sky to an observer on the surface of the Earth. A geosynchronous satellite in an inclined, circular orbit will sweep out a ground trace in the shape of a figure eight.
Ground track (trace)	The intersection of the Earth's surface with lines projected from the satellite to the Earth's center as the satellite moves in its orbit.
High Data Rate	Rates of 2,048 Mbps and higher.
Inclination	The angle between the plane of the orbit of a satellite and a reference plane. For a satellite in Earth orbit, inclination is the angle between the orbital plane and the equatorial plane as the satellite crosses the equator northward. Inclination values may be any angle between zero and 180 degrees.
Infrared radiation	Electromagnetic radiation with wavelengths between about 0.7 to 1000 micrometers. Infrared waves are not visible to the human eye. Longer infrared waves are called thermal infrared waves.
Ionization	The process by which a neutral atom, or a cluster of such atoms, becomes an *ion*. This may occur, for instance, by absorbtion of light ("photoionization") or by a collision with a fast particle ("impact ionization").
Ionosphere	A region covering the highest layers in the Earth's atmosphere, containing an appreciable population of *ions* and free *electrons*. The ions are created by sunlight ranging from the ultra-violet to x-rays.
Launch window	The period of time during which a satellite can be launched directly in to a specific orbital plane from a specific launch site.
Look angle	The angle between perpendicular and the direction an antenna or sensor is pointed.

Low Data Rate	Rates between 75bps and 2.4Kbps.
Magnetopause	The boundary of the magnetosphere, separating plasma attached to Earth from the one flowing with the solar wind.
Magnetic storm	A large-scale disturbance of the *magnetosphere,* usually initiated by the arrival of an *interplanetary shock,* originating on the Sun.
Major Axis	The distance form apogee to perigee measured through both foci. The longest diameter of an ellipse.
Medium Data Rate	Rates greater than 2.4 Kbps up to and including 2,048 Mbps.
Microsatellite	A satellite weighing between 10 – 100 kg.
Minisatellite	A satellite weighing between 100 – 500 kg.
Molniya orbit	A highly inclined (typically about 63.4 degrees), highly elliptical orbit with a 12-hour period.
Multispectral imagery	Imagery collected by a single sensor in multiple regions (bands) of the electromagnetic spectrum.
Nadir	The point on the Earth's surface directly below the satellite.
Nanosatellite	A satellite weighing between 1 – 10 kg.
Narrowband	Encompasses data rates less than 64 kbps.
National systems	A term used generically to refer to any asset used by the intelligence collection organizations of the US, especially space-based systems.
Negation	Measures to deceive, disrupt, deny, degrade or destroy an adversary's space systems and services
Orbital Period	The length of time (usually measured in minutes) for a satellite to complete one revolution.
Panchromatic	Black and white imagery that spans an area of the electromagnetic spectrum, typically the visual region.
Particle	In general, a charged component of an atom, that is, an ion or electron.

Particle beam	Stream of subatomic particles that are accelerated to high fractions of the speed of light and formed into non-divergent beams.
Passive sensor	One type of remote sensing instrument, a passive sensor picks up radiation reflected or emitted by the Earth. ETM+ is a passive remote sensing system
Payload	That portion of the load on the satellite for which a customer is willing to pay. Also, in general terms, the satellite to be delivered by the rocket.
Perigee	The point in a satellite's orbit where it is closest to Earth and its velocity is highest.
Period	The length of time it takes for the satellite to complete one orbit.
Perturbations	External forces that will cause a satellite to deviate from its normal orbital path.
Phased Array Radar	Any of a class of radars that, instead of using a rotating dish antenna to scan, uses electronically steered beams. The typical coverage of this fixed radar is 120 degrees per face or side; a phased array radar with three faces can provide coverage in all directions.
Picosatellite	A satellite weighing less than 1 kg.
Pixel	Picture element, the smallest element of a digital image.
Platform	A launched object, provides initial stabilization and orientation for payload or upper stage.
Polar orbit	An orbit with its plane aligned parallel with the polar axis of the Earth. Landsat has a polar orbit and flies 438 miles (705 kilometers) above the Earth.
Prevention	Measures to preclude an adversary's hostile use of third party or US space systems and services
Radar	Radio detection and ranging.
Radiation belt	The region of high-energy particles trapped in the Earth's magnetic field.
Radiometer	A device that detects and measures electromagnetic radiation.
Radiometric resolution	The ability of a sensor to detect levels of reflectance.

Regenerative Channel	On a regenerative channel, the satellite receiver converts the data signal (from the ground or airborne terminal) back into its original "ones" and "zeros." The satellite then uses those "ones" and "zeros" to retransmit the data toward the Earth to the receiving terminal.
Resolution	A unit of granularity in imagery.
Retrograde orbit	Any orbit with an inclination greater than 90 degrees.
Satellite	An object in space that is in orbit around another more massive object.
Scintillation	Random fluctuations in a transmitted signal's amplitude, phase or frequency.
Sensors	Electronic equipment used to find things. Sensors can be either active or passive.
Slot	That longitudinal position in the geosynchronous orbit into which a communications satellite is "parked."
Space control	Combat, combat support, and combat service support operations to ensure freedom of action in space for theUnited States and its allies and, when directed, deny an adversary freedom of action in space. The space control missionarea includes: surveillance of space; protection of US and friendly space systems; prevention of an adversary's ability to use space systems and services for purposes hostile to US national security interests; negation of space systems and services used for purposes hostile to US national security interests; and directly supporting battle management, command, control, communications, and intelligence. See also combat service support; combat support; negation; space; space systems.
Space Force Application	Combat operations in, through, and from space to influence the course and outcome of conflict. The space force application mission area includes ballistic missile defense and force projection. See also ballistic missile; force protection; space.
Space Force Enhancement	Combat support operations to improve the effectiveness of military forces as well as support other intelligence, civil, and commercial users. The space force enhancement mission area includes: intelligence, surveillance, and reconnaissance; integrated tactical warning and attack assessment; command, control, and communications; position, velocity, time, and navigation; and environmental monitoring.

Space environment	The region beginning at the lower boundary of the Earth's ionosphere (approximately 50 km) and extending outward that contains solid particles (asteroids and meteoroids), energetic charged particles (ions, protons, electrons, etc.), and electromagnetic and ionizing radiation (x-rays, extreme ultraviolet, gamma rays, etc.).
Space segment	That portion of a space system (see below) that is located in space, i.e., the satellite.
Space support	Combat service support operations to deploy and sustain military and intelligence systems in space. The spacesupport mission area includes launching and deploying space vehicles, maintaining and sustaining spacecraft on-orbit, anddeorbiting and recovering space vehicles, if required.
Space superiority	The degree of dominance in space of one force over another that permits the conduct of operations by the former and its related land, sea, air, space, and special operations forces at a given time and place without prohibitive interference by the opposing force
Space surveillance	The observation of space and of the activities occurring in space. This mission is normally accomplished with the aid of ground-based radars and electro-optical sensors. This term is separate and distinct from the intelligence collection mission conducted by space-based sensors which surveil terrestrial activity
Space system	An organization made up of equipment, some of which is in space, and people whose purpose is to perform specific technical tasks with the equipment. Space systems are almost universally made up of three principle subsystems, or segments; the space segment (satellite), the user segment (equipment and persons used to exploit the satellite's products), and the control segment (equipment and persons dedicated to maintaining the satellite).
Spatial resolution	The smallest sized feature that can be distinguished from surrounding features usually stated as a measure of distance on the ground and, in a digital image, directly associated with pixel size.
Space weather	The conditions and phenomena in space and specifically in the near-earth environment that may affect spaceassets or space operations. Space weather may impact spacecraft and ground-based systems. Space weather is influencedby phenomena such as solar flare activity,

	ionospheric variability, energetic particle events, and geophysical events..
Spectral resolution	The ability of a sensor to detect information in discrete regions of the electromagnetic spectrum.
Spot beam	A focused antenna pattern sent to a limited geographical area.
Spread spectrum	A technique used to overcome deliberate communications interference in which the modulated information is transmitted in a bandwidth considerably greater than the frequency content of the original information.
Solar wind	Hot solar *plasma* spreading from the Sun's *corona* in all directions, at a typical speed of 300-700 km/sec. It is caused by the great heat of the *corona*.
Space	The universe outside of the Earth's atmosphere. There is no universal definition of where space begins. For Army purposes, it is practical to define space as being the universe beyond the minimum altitude of a satellite in a circular orbit, about 89 miles.
Spoof	To cause a receiver to display, report, or cause to be carried out erroneous information or actions.
Sunspot cycle (or solar cycle)	An irregular cycle, averaging about 11 years in length, during which the number of sunspots (and of their associated outbursts) rises and then drops again.
Sun-synchronous orbit	An orbit in which a satellite is always in the same position with respect to the rotating Earth at the same time of day. The satellite travels around the Earth in the same direction, at an altitude of approximately 438 miles (705 kilometers).
Telemetry	Electronic remote monitoring of a launch vehicle's or satellite's functions. The purpose of telemetry is to provide ground control personnel with information as to the "health" and activity of a satellite.
Temporal resolution	The space of time between collection of successive images.
Thermal infrared	Electromagnetic radiation with wavelengths between 3 and 25 micrometers.

Transfer Orbit	A highly elliptical orbit which is used as an intermediate stage for placing satellites into their final orbit.
Uplink	The Earth-to-space telecommunications pathway
User Segment	That portion of the space system that is ground-based and provides useful products. This may consist of receivers, processors, and special support personnel at a fixed site, or a simple portable radio that provides satellite access.
Visible radiation	The electromagnetic radiation that humans can see as colors. The visible spectrum is made up of wavelengths between 0.4 to 0.7 micrometers. Red is the longest and violet is the shortest.
Wavelength	Wavelength is literally the length of one complete cycle (wave) of an electromagnetic signal. In radio signal terms, for example this can be determined by dividing the speed of light in meters per second (about 300,000,000) by the frequency, measured in hertz (also known as cycles per second). The result will be wavelength in meters.
Wideband	Encompasses data rates greater than 64 kbps.